VACUUM

SCIENCE,
TECHNOLOGY
AND
APPLICATIONS

VACUUM
SCIENCE, TECHNOLOGY AND APPLICATIONS

Pramod K. Naik

CRC Press is an imprint of the
Taylor & Francis Group, an **informa** business

CRC Press
Taylor & Francis Group
6000 Broken Sound Parkway NW, Suite 300
Boca Raton, FL 33487-2742

First issued in paperback 2020

ISBN-13: 978-0-367-57206-8 (pbk)
ISBN-13: 978-1-138-58715-1 (hbk)

Visit the Taylor & Francis Web site at
http://www.taylorandfrancis.com

and the CRC Press Web site at
http://www.crcpress.com

Contents

About the Author

Pramod K. Naik has been actively involved in Vacuum Science & Technology since 1958. He obtained Ph. D. in Physics from the University of Bombay (now Mumbai University). He investigated electrical oscillations in “Philips Ionization Gauge” and allied type of D.C. glow discharge with Dr. V T Chiplonkar at the Institute of Science, Mumbai. He joined the Atomic Energy Establishment, Trombay (now Bhabha Atomic Research Centre) at Mumbai in 1961. At BARC, he was engaged in development of high and ultrahigh vacuum components and systems. He was appointed as a member of the expert group formed for the Synchrotron Radiation Source.

At BARC, he developed thermal desorption spectrometer and investigated entrapment of energetic inert gas ions near molybdenum solid surface. He has published many papers in reputed international journals.

He worked with Dr. R.G Herb at the University of Wisconsin, Madison, Wisconsin USA in the field of ultrahigh vacuum and developed the glass orbitron pump.

He has also worked at the Westinghouse Electric Corporation (now Eaton Corporation) where he was engaged in absorption of the vacuum interrupter technology and was a consultant to the Eaton Corporation (USA) during 1996-2016. He worked as a manager at Crompton Greaves, Aurangabad.

He has been associated with the Indian Vacuum Society and has organized a number of short-term courses in Vacuum Science & Technology. He received the Lifetime Achievement Award for his outstanding contributions in the field of Vacuum Science & Technology from the Indian Vacuum Society in the year 2003.

Preface and Acknowledgements

When I joined a team of research students at the Institute of Science, Mumbai (formerly known as the Royal Institute of Science) in the late 1950's, the team was engaged in the experimental studies of plasma and electrical discharges in vacuum under the guidance of Dr. V. T. Chiplonkar. At this stage, I had my first interaction with vacuum. The laboratories were then equipped with rotary pumps, diffusion pumps, manometers and Pirani gauges. I was conversant with constructing glass vacuum systems using elementary glass-blowing techniques. The appearance of colorful glow discharges in tubes under vacuum conditions in the laboratories used to attract many curious onlookers.

Later, upon joining the Atomic Energy Establishment (now Bhabha Atomic Research Centre) at the Tata Institute of Fundamental Research at Mumbai in 1961, I was further exposed to a wide variety of vacuum components and systems. Mr. C. Ambasankaran, who had earlier worked with Metropolitan Vickers, UK was then engaged in a program of development of vacuum technology at the Atomic Energy division. Dr. Homi Bhabha who was the Chairman of the Atomic Energy Commission had planned to introduce vacuum technology in India, realizing its important potential applications in many fields in science and industry.

I had an opportunity of witnessing the growth of vacuum technology in India. Some of the noteworthy contributions made by the team during my sojourn at the Atomic Energy department were the indigenous developments of oil diffusion pumps, sputter-ion and orbitron pumps, cryo-sorption pumps, ion gauges, vacuum hardware, residual gas analyzers/mass spectrometers, thin film coating units, surface analytical equipment including Auger electron spectrometer, thermal desorption spectrometer, field electron/field ion microscope. Vacuum systems for electron beam welding, freeze-drying, space simulation studies were also developed in addition to design and construction of large vacuum systems for cyclotron accelerator and synchrotron radiation source.

During mid 1960's, I was engaged in development of the orbitron ion pump with Dr. R. G. Herb at the University of Wisconsin, USA at Madison where I was exposed to ultrahigh vacuum techniques.

The Indian Vacuum Society (IVS) was formed in 1970. IVS has been an excellent platform for the workers from research organizations, universities and industries, in the fields associated with vacuum technology to exchange information. I was actively engaged in conducting many short-term courses on vacuum science & technology organized by IVS as I covered most topics in the subject during the courses. The lecture notes that I had prepared for these courses have been much useful for this book.

Later in mid-1980's, I had an opportunity to work at the Vacuum Interrupter plant of Westinghouse Electric Corporation USA (now Eaton Corporation) as I was involved in absorption of technology of Vacuum Interrupters. It was interesting and rewarding experience to observe utilization of ceramic metalizing and vacuum brazing techniques for large scale manufacturing of vacuum interrupters and then implementing the technology in India.

The knowledge and the experience gained at various stages have prompted me to write this book which might serve as a small contribution to the wealth of information on the subject that already exists in form of excellent books and articles written by stalwarts in the field.

I am indebted to Late Dr. V.T. Chiplonkar who introduced me to the world of vacuum technology in 1958. Also, I must acknowledge the valuable guidance provided by Late Dr. Ray Herb in specializing ultrahigh vacuum technology.

The encouragement and help that I received from my wife Sharada, sons Hrishikesh and Parijat is appreciated. Hrishikesh has carefully reviewed the contents including the illustrations in the book.

I am thankful to my friends and colleagues who have helped me with their support in writing this book. Dr. Milind Bemalkhedkar and Dr. Shriram Mhaskar were kind to review the scientific content of the book. I appreciate the help received from Mr. Girish Magre for getting the book printed and published. I do appreciate the permissions given by the individuals and institutions for printing their copyright images in this book. My apologies in those few cases where I was unable to approach the concerned to obtain the permission. Such cases may please be brought to my attention so that I could approach the appropriate authority for permission.

I am indebted to a number of scientists and engineers who have made immense contribution to the subject in the form of informative books and publications. These include: Drs. S. Dushman and J. Lafferty, Dr. D. Alpert, Dr. R. Herb, Dr. G. Lewin, Dr. M. Kaminsky, Dr. J. Cobine, Dr. F. Rosebury, Dr. P. Slade, Dr. P. Redhead, Dr. P. Hobson and Dr. E. Korenelsen.

Introduction

Since the presence of vacuum was first identified in 1644 by Torricelli in his famous experiment with a glass tube filled with mercury and a mercury reservoir, many illustrious workers have contributed to the advancement of vacuum science and technology. Theories presented by the scientists have helped in understanding of concepts of the kinetic theory of gases which form the basis of vacuum science. The development of the key vacuum devices such as gauges, pumps was possible due to the valuable and ingenious work by many workers. Innumerable physical phenomena were investigated using applications and inventions based on vacuum technology.

Ideally, vacuum is a space that is empty of matter. Absolute vacuum is a theoretical concept. The universal free space offers a situation closest to absolute vacuum with the gas density of a few hydrogen atoms per cubic centimeter equivalent to a pressure of about 10^{-14} Pascal at room temperature. Practically, vacuum is a region with a gas pressure much less than the atmospheric pressure.

Vacuum has an important role in science and technology. Study of the interaction of charged particles, neutrals and radiation with each other and with solid surfaces requires a vacuum environment for reliable investigations. Vacuum has contributed immensely to advancements made in nuclear science, space, metallurgy, electrical/ electronic technology, chemical engineering, transportation, robotics and many other fields.

This book is intended to assist students, scientists, technicians and engineers to understand the basics of vacuum science and technology for application in their projects.

1

Kinetic Theory of Gases

1.1 Basic Concepts

For most calculations of vacuum parameters, it is necessary to use the concepts of the kinetic theory of gases. The discussion in this chapter is limited to the relevance of these concepts to vacuum science and technology. The reader is suggested also to refer to the excellent literature[1,2,3,4] published on the subject. The basic postulates of the kinetic theory of gases are

- Molecules of a gas in a volume are in constant random motion. The motion is related to the temperature of the gas
- Molecules collide elastically with each other and with the container wall
- Gas constitutes a large number of molecules. All the molecules of a given chemical substance are exactly alike
- The molecules are separated by distances, much larger in comparison with their own dimensions
- The molecules exert no force on each other or on the walls of their container except when they collide.

The theory based on these assumptions satisfactorily explains the behaviour of real gases.

1.1.1 Pressure

The pressure of a gas at a certain point in a given direction is defined as the temporal rate of momentum transfer in the assigned direction across a unit area normal to that direction. Pressure P is produced by the impulse of the gas molecules hitting the wall of the container. Table 1.1 shows conversions of different commonly used units of pressure.

Table 1.1: Conversions of Units of Pressure

	Pascal (Pa)	Torr	Standard atmosphere	Millibar	Dynes · cm^{-2}
1 Pascal (N · m^{-2})	1	7.5×10^{-3}	9.90×10^{-6}	10^{-2}	10
1 Torr	133.28	1	1.31×10^{-3}	1.33	1.33×10^{3}
1 Standard atmosphere	1.01×10^{5}	760	1	1.01×10^{3}	1.01×10^{6}
1 Millibar (mbar)	10^{2}	0.75	9.90×10^{-4}	1	1×10^{3}
1 Dynes · cm^{-2}	10^{-1}	7.5×10^{-4}	9.90×10^{-7}	10^{-3}	1

1.1.2 Equation of State

For an ideal gas,

$$PV = NkT \tag{1.1}$$

Also,

$$P = nkT \tag{1.2}$$

where P – pressure of the gas; V – volume; N – the number of molecules; n – the number of molecules in the unit volume (gas density); k – Boltzmann constant = 1.38×10^{-16} erg · K^{-1}; T – the temperature of gas in Kelvin.

Useful conversion of units for vacuum calculations:

Amount of gas: 1 Torr · liter = 1.33×10^{-1} Pa·m^3; outgassing rate/gas impingement rate: 1 Torr · liter · s^{-1} · cm^{-2} = 1.33×10^{3} Pa · m · s^{-1}; Boltzmann constant $k = 1.38 \times 10^{-16}$ erg · K^{-1}; = 1.03×10^{-22} Torr · liters · K^{-1}, if the pressure is in Torr; = 1.37×10^{-23} Pa · m^3 · K^{-1}, if the pressure is in Pa.

1.1.3. Gas Density

From equation (1.2), the gas density n is given by

$$n = \frac{P}{kT}$$

The number of molecules in one liter at 1 Torr at 293 K is 3.3×10^{19}; the number of molecules in one liter at 1 Pa at 293 K is 2.48×10^{17};

the number of molecules in one m^3 at 1 Pa at 293 K is 2.48×10^{20}.

It is interesting to note that at relatively high temperatures, the gas pressure can be much higher although the gas density is relatively much lower and equivalent to lower gas pressure at room temperature. This is the case in a thermonuclear reactor in which the gas pressure is close to the atmospheric pressure (1×10^5 Pa) and the temperature of about 10^8 K while the gas density is about 10^{20} atoms per m^3, equivalent to the gas pressure of about 4×10^{-1} Pa at room temperature.

1.1.4 Avogadro's Law

Avogadro's law states that equal volumes of gases at the same temperature and pressure contain the same number of molecules. The number of molecules for any gas in a volume of 22.41 liters at 760 Torr (1.01×10^5 Pa) and 0°C equals 6.023×10^{23}. This number is called Avogadro's number A_m. One mole of gas contains molecules equal to Avogadro's number.

If m is the mass of a molecule, then the gram molecular mass (weight)

$$M = mA_m$$

where M is amount of a molecular substance whose weight, in grams, is numerically equal to the molecular weight of that substance. One gram–molecular weight of molecular oxygen, O_2 (molecular weight approximately 32), is 32 grams, and one gram–molecular weight of water, H_2O (molecular weight approximately 18) is 18 grams. The molecular weight of an element or compound is expressed in grams (g). The molecular weight on a scale on which the atomic weight of the ^{12}C isotope of carbon is taken as 12. In the International System of Units, the gram–molecular weight is replaced by the mole.

The equation of state can be expressed in terms of moles as given below.

$$PV = n_m RT \tag{1.3}$$

where P and V are the pressure and the volume of the gas respectively, n_m equal to N/A_m is the number of moles and R is the gas constant equal to kA_m. In terms of vacuum units $R = 8.31451$ Pa · m^3 · K^{-1} · $mole^{-1}$. R is expressed in with other units as: 62.36 Torr · liter · K^{-1}·$mole^{-1}$; 8.31 Pa · m^3 · K^{-1}·$mole^{-1}$; 1.98 cal · K^{-1} · $mole^{-1}$.

PV corresponds to the translational kinetic energy of the gas.

1.1.5 Molecular Motion and Energy

Pressure P is produced by the impulse of the molecules hitting the wall:

$$P = \frac{1}{3} nmv_{rms}^2 \tag{1.4}$$

where m is the mass of a molecule; n is gas density; v_{rms} is the root mean square velocity of the molecules

The translational kinetic energy of a molecule is given by

$$\frac{1}{2} mv_{rms}^2 = \frac{3}{2}(kT) \tag{1.5}$$

$$v_{rms} = \left(\frac{3kT}{m}\right)^{\frac{1}{2}} = 15800\left(\frac{T}{M}\right)^{\frac{1}{2}} \text{cm} \cdot \text{s}^{-1} \tag{1.6}$$

where k is the Boltzmann constant expressed in erg · K^{-1}, T is the temperature of the gas in Kelvin, and M is the molecular weight.

The average velocity v_a is given by

$$v_a = \left(\frac{8kT}{\pi m}\right)^{\frac{1}{2}} = 14551\left(\frac{T}{M}\right)^{\frac{1}{2}} \tag{1.7}$$

1.1.6 Molecular Impingement Rate

The number ν of molecules incident on unit area per second is given by

$$\nu = \frac{1}{4} nv_a = \frac{n}{4}\left(\frac{8kT}{\pi m}\right)^{\frac{1}{2}} \tag{1.8}$$

$$= P(2\pi mkT)^{-\frac{1}{2}} \tag{1.9}$$

where P is the pressure of the gas; m is the mass of a molecule; k is the Boltzmann constant; n is the gas density; T is the temperature of gas in Kelvin; M is the molecular weight; v_a is the average velocity of gas

$$\nu = \frac{14.55\times10^{3}}{4} n\left(\frac{T}{M}\right)^{\frac{1}{2}} \quad \text{in cgs units}$$

$$= 3.46\times10^{22} P(MT)^{-\frac{1}{2}} \quad \text{if } P \text{ is in Torr}$$

$$= 2.63\times10^{24} P(MT)^{-\frac{1}{2}} \quad \text{if } P \text{ is in Pa} \tag{1.10}$$

1.1.7 Mean Free Path

The mean free path (MFP) is the average distance travelled by a molecule between two successive intermolecular collisions. It is given by

$$\lambda = \frac{kT}{\pi P\delta^2 \sqrt{2}} \tag{1.11}$$

where P is the pressure of the gas; k is the Boltzmann constant; T is the temperature of the gas; δ is the gas kinetic diameter of the molecule.

For a given gas temperature, the MFP λ is longer at lower pressure. At atmospheric pressure (1×10^5 Pa) and at room temperature, nitrogen has MFP of 6.3×10^{-6} cm while at a pressure of 1×10^{-7} Pa, the MFP is 6.3×10^{6} cm at the same temperature.

Table 1.2 shows gas kinetic data for nitrogen at different degrees of vacuum.

1.1.8 Heat Transfer

1.1.8.1 Heat Transfer by Radiation

The Stefan–Boltzmann law is expressed as

$$W = \sigma\varepsilon T^4 \tag{1.12}$$

where σ is the Stefan–Boltzmann constant (5.67×10^{-12} W·cm^{-2}·K^{-4}); ε is the ratio of the emissivity of the surface to the emissivity of a black body ($0 \leq \varepsilon \leq 1$); T is the temperature of gas in Kelvin.

Table 1.2. Gas kinetic data for nitrogen at room temperature

Condition	Pressure (Torr)	Pressure (Pascal)	Gas Density Molecules (cm^{-3})	Mean free path of gas* λ (cm)	Molecular impingement rate ($cm^{-2} \cdot s^{-1}$)	Molecular formation time** (s)
Atmospheric pressure	760	1.01×10^{5}	2.5×10^{19}	6.3×10^{-6}	2.9×10^{23}	2.6×10^{-9}
Vacuum	1	133.28	3.3×10^{16}	4.8×10^{-3}	3.9×10^{20}	2×10^{-6}
High vacuum	10^{-6}	1.33×10^{-4}	3.3×10^{10}	4.8×10^{3}	3.9×10^{14}	2
Ultrahigh vacuum	10^{-9}	1.33×10^{-7}	3.3×10^{7}	4.8×10^{6}	3.9×10^{11}	2×10^{3}
Extreme high vacuum	10^{-12}	1.33×10^{-10}	3.3×10^{4}	4.8×10^{9}	3.9×10^{8}	2×10^{6}

*Mean free path for electrons $\lambda_e = 4 \times \sqrt{2}\,\lambda$ and mean free path for ions $\lambda_i = \sqrt{2}\,\lambda$; **Assuming unit sticking probability and monolayer consisting of 7.8×10^{14} molecules·cm^{-2}

The heat transfer by radiation is predominant even at high pressures due to the T^4 dependence.

1.1.8.2 Heat Transfer by Conduction at Low Pressure

The molecular collisions with the wall are predominant at lower pressures. Consider that molecules of temperature T_i impinge on a wall of temperature T_w, where $T_w > T_i$. If T_r is the temperature of the molecules leaving the surface, the cooling rate will be proportional to the impingement rate υ of the gas molecules on the wall and the temperature difference $(T_w - T_i)$. The accommodation coefficient α, introduced by Knudsen[5], is given by

$$\alpha = \frac{T_r - T_i}{T_w - T_i} \tag{1.13}$$

A Maxwellian distribution is assumed here for the reflected molecules. The value of α depends on surface conditions[6].

The thermal conductivity κ of gas is defined as the heat flux per unit time across an area of unit cross section for the unit temperature gradient and is expressed in watts per square centimeter for a temperature gradient of 1°C per centimeter.

W = Impingement rate × transfer of translatory kinetic energy:

$$W = \frac{3}{2}k\nu\ (T_r - T_i) = \frac{3}{2}k\nu\alpha(T_w - T_i) = \frac{z\alpha P(T_w - T_i)}{(T_i M)^{1/2}} \tag{1.14}$$

where M is the molecular weight of the gas and z is a constant.

This establishes that the cooling rate is proportional to the pressure. The pressure measurement gauges based on molecular heat conduction utilize this dependence of the cooling rate on pressure. The transfer of kinetic energy by collisions on the microscopic scale is involved in the conduction of heat in this case.

The gas flow in the viscous flow range is influenced by the viscosity of the gas. In the viscous flow range, if the two adjacent layers of gas move with different velocities, different momenta are experienced by the molecules that cross the boundary between the layers. Viscosity results from the momentum transfer. The tangential force exerted per unit area at the unit velocity gradient is defined as the coefficient of viscosity and is given by

$$\eta = 0.5\lambda nmv_a \tag{1.15}$$

where λ is the mean free path of the gas; n is gas density; m is the mass of a molecule; v_a is the average velocity of the gas.

The cgs unit of η is poise (1 poise = 1 dyne·s·cm^{-2}). It is established that the thermal conductivity κ is a function of the coefficient of viscosity η and the specific heat at constant volume c_v, and is given by

$$\kappa = 2.5\eta c_v \tag{1.16}$$

for monatomic gases. The constant factor is less than 2.5 for polyatomic gases. In the viscous flow range, the viscosity and the heat conductivity are functions of temperature only and are independent of pressure provided the density is not too high and the intermolecular attractive forces are not involved.

1.1.8.3 Convection

If a temperature gradient exists in the direction of gravitational force, the heat is transferred by convection by the macroscopic upward flow of gas resulting from the difference in gas density.

1.1.9 Thermal Transpiration

Consider two chambers maintained at different temperatures T_1 and T_2 and pressures P_1 and P_2 respectively, interconnected by a tube as shown in Figure 1.1.

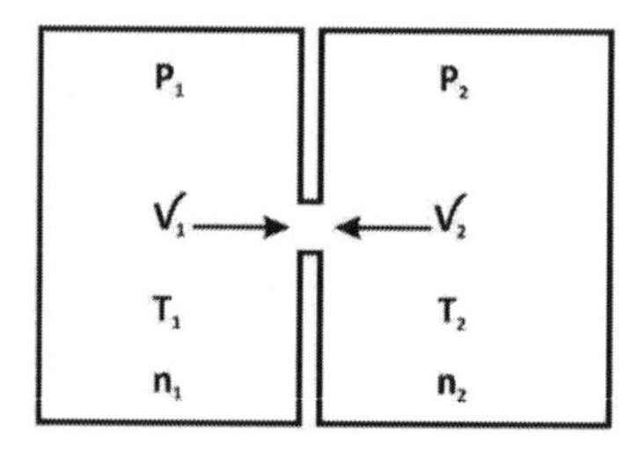

Fig. 1.1. Thermal transpiration between two chambers separated by a partition with a tube.

In this situation, thermal transpiration will occur if the pressure is low enough to facilitate the molecular flow. Equilibrium demands that the impingement rate of the gas is equal in either direction of the tube. Thus, using equation (1.9), we have

$$\nu_1 = \nu_2$$

where ν_1 , ν_2 are the gas impingement rates in the opposite directions and n_1 and n_2 are the corresponding gas densities. Therefore,

$$\frac{P_1}{(2\pi mkT_1)^{\frac{1}{2}}} = \frac{P_2}{(2\pi mkT_2)^{\frac{1}{2}}}$$

$$\left(\frac{P_1}{P_2}\right) = \left(\frac{T_1}{T_2}\right)^{\frac{1}{2}}$$

$$\left(\frac{n_1 kT_1}{n_2 kT_2}\right) = \left(\frac{T_1}{T_2}\right)^{\frac{1}{2}}$$

$$\left(\frac{n_1}{n_2}\right) = \left(\frac{T_2}{T_1}\right)^{\frac{1}{2}} \tag{1.17}$$

It is necessary to consider thermal transpiration if the gas density is measured in a region having a temperature that differs from the temperature of the gauge as in the case of a vacuum furnace. It is interesting to observe that the pressure in the hot zone of a vacuum furnace is higher and the gas density is lower than in the gauge.

References

1. E. H. Kennard, "Kinetic Theory of Gases", McGraw Hill Book Company, New York (1938).
2. S. Dushman and J. M. Lafferty, "Scientific Foundation of Vacuum Technique" Second Edition, John Wiley and Sons, Inc., New York (1962).
3. G. Lewin, "Fundamentals of Vacuum Science and Technology", McGraw Hill Book Company, New York (1965).
4. A. Guthrie and R. Wakerling, "Vacuum Equipment and Techniques", McGraw Hill Book Company, New York, (1949).
5. M. Knudsen, Ann. der Physik **34**, 593 (1911).
6. H. Y. Wachman, American Rocket Society Journal **32**, No. 1, January (1962).

2

Flow of Gas and Conductance

2.1 Types of Gas Flow

Types of gas flow encountered in vacuum systems include

- Viscous flow
- Molecular flow
- Transition (between viscous and molecular) flow

The type of gas flow is generally decided by the value of a dimensionless parameter called the Knudsen number K_n which is the ratio of the MFP λ of the molecule to a characteristic dimension D of the vessel through which the gas flows, such as the diameter of a tube and can be expressed as

$$K_n = \frac{\lambda}{D} \tag{2.1}$$

In case of viscous flow, the value of K_n is small as the MFP λ is much smaller than the characteristic dimension of the vacuum chamber and the inter-molecular collisions predominate the molecular collisions with walls.

2.1.1 Viscous Flow

The viscous flow is divided into turbulent and laminar flows. High pressure gradients can cause the turbulent flow which is characterized by orderless swirling eddies of the gas. In this flow, velocities and directions of the gas particles may differ with the average velocity and the overall direction of the gas flow.

The laminar flow, characterized by absence of turbulence, is smoother and orderly in direction of flow. In the laminar gas flow inside a tube of circular cross section, the motion of gas is in the axial direction with the reduction of the gas velocity near the wall.

The Poiseuille equation of the steady laminar flow through tubes is given by

$$Q = \frac{\pi R^4 \Delta P}{8\eta L} \quad \text{in cgs units} \tag{2.2}$$

where Q is the volumetric flow rate; R is the radius of the tube; L is the length of the tube; ΔP is the pressure drop in the direction of flow; η is the specific viscosity.

The Poiseuille equation assumes that

- The gas is not compressible
- The flow velocity profile is constant across the length
- Turbulent motion is absent in the gas
- Flow velocity at the tube wall is zero.

In this case, the gas flow can be considered similar to the flow of the fluid involving the coefficient of viscosity. The viscous flow is characterized by the adjacent layers of gas moving with different velocities. The coefficient of viscosity η has been defined as the tangential force per unit area exerted at unit velocity gradient and is given by the equation 1.15 as

$$\eta = 0.5\lambda n m v_a \quad \text{in cgs units}\left(\text{dynes} \cdot \text{s} \cdot \text{cm}^{-2}\right)$$

The viscous flow of gases (and fluids) is also characterized by the Reynolds number R_e which is the ratio of inertial forces to viscous forces. R_e is associated with the density and viscosity of the gas/fluid, along with velocity and a characteristic dimension. The Reynolds number is defined as

$$R_e = \frac{\rho u L}{\mu} = \frac{uL}{\nu} \quad \text{in cgs units} \tag{2.3}$$

where u is the mean velocity; L is the characteristic linear dimension; μ is the dynamic viscosity of the fluid; ν is the kinematic viscosity $= \mu/\rho$; ρ is the density of the fluid/gas.

For flow in a pipe or tube, the Reynolds number can be expressed as

$$R_e = \frac{\rho u D}{\mu} = \frac{uD}{\nu} = \frac{QD}{\nu A} \quad \text{in cgs units} \tag{2.4}$$

where D is the diameter of the pipe; Q is the volumetric flow rate; A is the pipe cross-sectional area; u is the mean velocity of the fluid/gas; μ is the dynamic viscosity of the fluid/gas; v is the kinematic viscosity.

R_e is also used to characterize different flow regimes, such as laminar or turbulent flow. In general, $R > 2200$ for turbulent flow; $R < 1200$ for laminar flow; $1200 < R < 2200$ turbulent or viscous flow depending on the geometry.

2.1.2 *Molecular Flow*

The molecular flow takes place at larger values of the Knudsen number K_n when MFP λ is much larger than the characteristic dimension D of the vacuum chamber and the inter-molecular collisions are comparatively few as compared to the molecular collisions with walls. Molecules move randomly and freely, independent of each other in the molecular flow.

The transition flow is an intermediate state wherein both types of collisions (intermolecular and those with walls) govern the flow. Generally, $K_n < 0.01$ for viscous flow; $1 > K_n > 0.01$ for transition flow and $K_n > 1.0$ for molecular flow.

The rate of flow Q of gas across an area can be defined as

$$Q = \frac{PdV}{dt} \tag{2.5}$$

where P and V are the pressure and the volume of the gas. As PV is considered as translational energy, Q can be considered as the rate of flow of energy. The number of molecules moving across the area is proportional to Q. The rate of the mass flow can be expressed as

$$n\frac{dV}{dt} = Q(kT)^{-1} \tag{2.6}$$

where k is the Boltzmann constant and T is the temperature of the gas.

2.2 Conductance

The conductance C of a component is defined as the ratio of the flow rate Q to the pressure difference ΔP

$$C = \frac{Q}{\Delta P} \tag{2.7}$$

$$C = C_1 + C_2 \quad \text{for parallel connection} \tag{2.8}$$

and

$$\frac{1}{C} = \frac{1}{C_1} + \frac{1}{C_2} \quad \text{for series connection} \tag{2.9}$$

C_1 and C_2 are the conductances of components used in the connection.

2.2.1 Conductance of an Orifice in Molecular Flow Region

The flow rate Q is given by

$$Q = P\frac{dV}{dt} = kT\frac{dN}{dt}$$

where P and V are the pressure and the volume of the gas respectively, k is the Boltzmann constant and T is the temperature of the gas in Kelvin.

Since the flow rate Q in one direction is (kT) × (impingement rate of gas) × (area of the orifice), using equation (1.8), we have

$$Q = kT\left(\frac{Anv_a}{4}\right) = kT\left[PA(2\pi mkT)^{-\frac{1}{2}}\right] = AP\left(\frac{kT}{2\pi m}\right)^{\frac{1}{2}}$$

where A is the area of the orifice; n is the gas density; v_a is the average velocity of the gas; m is the mass of the molecule; T is the temperature of gas in Kelvin.

The net flow downstream Q is expressed on replacing P by ΔP in the equation

$$Q = A\Delta P\left(\frac{kT}{2\pi m}\right)^{\frac{1}{2}}$$

Thus conductance C_{mo} of the orifice is given by

$$C_{mo} = \frac{Q}{\Delta P} = A\left(\frac{kT}{2\pi m}\right)^{\frac{1}{2}} \quad \text{in cgs units} \tag{2.10}$$

$$= 3.64A\left(\frac{T}{M}\right)^{\frac{1}{2}} \quad \text{liters} \cdot \text{s}^{-1}$$

if A is expressed in cm^2

$$= 36.4A\left(\frac{T}{M}\right)^{\frac{1}{2}} \quad \text{m}^3 \cdot \text{s}^{-1} \tag{2.11}$$

if A is expressed in m^2.

The value of conductance calculated from the equation above needs a correction if the vessel dimensions are comparable to the area A of the aperture. In such cases, if A_0 is the cross sectional area of the tube, a multiplication factor $A_0/(A_0-A)$ is used with C_{mo} to obtain the effective value of the conductance.

Conductance C_{mo} of an orifice for nitrogen in the molecular flow region at 293 K is

$$C_{moN_2} = 11.8 \text{ A liters} \cdot \text{s}^{-1}$$

if A is in cm^2

$$= 118 \text{ A m}^3 \cdot \text{s}^{-1}$$

if A is in m^2.

Conductance of a Long Tube

To derive the conductance C_{mlt} of a long tube of diameter $D \ll$ length L in the molecular flow region, it is necessary to consider

- the force that drives the gas through the tube. This equals the product of pressure difference and the cross sectional area.
- the frictional force transferred by the gas to the tube wall. This is equivalent to the product of the gas impingement rate on the surface area of the tube wall and the net momentum in the axial direction.

Using equation (1.8), the force F is given by

$$F = \frac{\Delta P \pi D^2}{4} = \frac{(\pi DLmu)nv_a}{4} \tag{2.12}$$

where u is the net axial gas velocity; m is the mass of the gas molecule.

Thus,

$$\Delta P = \frac{n v_a L m u}{D} \tag{2.13}$$

Also, we can consider

$$Q = F \times u \tag{2.14}$$

Substituting value of F from equation (2.12), we get

$$Q = \frac{u \Delta \mathrm{P} \pi \mathrm{D}^2}{4}$$

As $P = nkT$ from equation (1.2),

$$Q = \frac{nkTu\pi D^2}{4} \tag{2.15}$$

As $C = \dfrac{Q}{\Delta P}$

Substituting values of Q and ΔP from equations (2.15) and (2.13), we get

$$C_{mlt} = \left(\frac{nkTu\pi D^2}{4}\right) \times \left(\frac{n v_a L m u}{D}\right)^{-1}$$

Substituting v_a by $\left(\dfrac{8kT}{\pi m}\right)^{\frac{1}{2}}$ from equation (1.7)

$$C_{mlt} = \left(\frac{\pi D^3}{16L}\right) \times \left(\frac{2\pi kT}{m}\right)^{\frac{1}{2}} \tag{2.16}$$

Rigorous derivation gives

$$C_{mlt} = \left(\frac{2\pi kT}{m}\right)^{\frac{1}{2}} \times \left(\frac{D^3}{6L}\right) \quad \text{(in cgs units)} \tag{2.17}$$

and

$$C_{mlt} = 3.8\left(\frac{D^3}{L}\right) \times \left(\frac{T}{M}\right)^{\frac{1}{2}} \text{ liters} \cdot \text{s}^{-1}$$

if D and L are in cm

$$C_{mlt} = 38\left(\frac{D^3}{L}\right)\times\left(\frac{T}{M}\right)^{\frac{1}{2}}\ \mathrm{m^3\cdot s^{-1}} \tag{2.18}$$

if D and L are expressed in meters.

Thus for nitrogen at 293 K, the molecular conductance C_{mltN_2} is given by

$$C_{mltN_2} = 123\left(\frac{D^3}{L}\right)\ \mathrm{m^3\cdot s^{-1}}$$

with D, L in m, and

$$C_{mltN_2} = 12.3\left(\frac{D^3}{L}\right)\ \mathrm{liters\cdot s^{-1}}$$

with D, L in cm.

A cold trap filled with liquid nitrogen has a conductance of about 50% of its conductance at room temperature.

For a short tube, the conductance C_{ap} of the aperture at the entrance needs to be considered. Thus the effective conductance C_{emst} is given by

$$\frac{1}{C_{emst}} = \frac{1}{C_{mlt}} + \frac{1}{C_{ap}}$$

It can be shown that

$$C_{emst} = C_m\left(1+\frac{4D}{3L}\right)^{-1} \tag{2.19}$$

2.2.2 *Conductance in the Viscous Flow Region*

Conductance of an Orifice

As the inter-molecular collisions of gas particles dominate over the collisions of the gas with the walls, the wall effects due to viscosity can be ignored. The gas can be considered as a liquid in the viscous flow. In the viscous flow region, the gas flow is governed by the ratio $r = P_2/P_1$ of the downstream to the upstream pressure. The flow is independent of P_2 as the gas flows with the sonic velocity at low pressure ratios. Consider an orifice to be thin and having a sharp edge. The viscous flow rate Q is given by[1]

$$Q = AP_1\left[\frac{2\gamma kT}{m(\gamma-1)}\right]^{\frac{1}{2}} r^{\frac{1}{\gamma}}\left[1-r^{\left(\frac{\gamma-1}{\gamma}\right)}\right]^{\frac{1}{2}} \text{(in cgs units)} \qquad (2.20)$$

where k is the Boltzmann constant; m is the mass of the gas molecule; T is the temperature of the gas in Kelvin; A is the annular area of the orifice; γ is the ratio c_p/c_v of specific heats of the gas (which is 1.4 for diatomic species and 1.667 for monatomic species).

Thus, for air at 293 K, we have

For $r \geq 0.52$,

$$C_{vo} = 76.6r^{0.712}\left(1-r^{0.288}\right)^{\frac{1}{2}}\frac{A}{1-r} \quad \text{liters} \cdot \text{s}^{-1}$$

if A is in cm^2 and

$$C_{vo} = 766r^{0.712}\left(1-r^{0.288}\right)^{\frac{1}{2}}\frac{A}{1-r} \quad \text{m}^3 \cdot \text{s}^{-1} \qquad (2.21)$$

if A is in m^2.

For $r \leq 0.52$,

$$C_{vo} \approx 20\frac{A}{1-r} \quad \text{liters} \cdot \text{s}^{-1}$$

if A is in cm^2 and

$$C_{vo} \approx 200\frac{A}{1-r}\text{m}^3 \cdot \text{s}^{-1} \qquad (2.22)$$

if A is in m^2

For $r \leq 0.1$,

$$C_{vo} \approx 20A \quad \text{liters} \cdot s^{-1}$$

if A is in cm^2 and

$$C_{vo} \approx 200A \quad \text{m}^3 \cdot \text{s}^{-1} \qquad (2.23)$$

if A is in m^2.

Conductance of a Tube

The flow lines are parallel in case of the laminar flow and the velocity of the gas near the walls is reduced by the viscosity. It is assumed that the gas is incompressible. This is possible if the flow rate Q is as given below

$$Q < \frac{\pi D^2 P v_s}{12} \times 10^{-3} \quad \text{Torr} \cdot \text{liters} \cdot \text{s}^{-1}$$

where v_s is the sonic velocity 3.44×10^4 cm/s for air at 20°C, D is the diameter in cm and P is the pressure in Torr. Or if

$$Q < \frac{\pi D^2 P v_s}{12} \quad \text{Pa} \cdot \text{m}^3 \cdot \text{s}^{-1} \tag{2.24}$$

where v_s is the sonic velocity 3.44×10^2 m/s for air at 20°C, D is the diameter in m and P is the pressure in Pa.

The viscous conductance C_{vlt} of a long tube is given by

$$C_{vlt} = \frac{\pi}{128} \frac{D^4 P_a}{\eta L} \quad (\text{in cgs units}) \tag{2.25}$$

where D and L are the diameter and the length of the tube, P_a is the average pressure in the tube and η stands for the viscosity of the gas.

Considering $\eta = 175 \times 10^{-6}$ Poise for nitrogen at 293 K, conductance C_{vltN_2} of a long tube for nitrogen at 293 K in the viscous flow region is

$$C_{vltN_2} = 1.88 \times 10^2 \frac{D^4 P_a}{L} \text{ liters} \cdot \text{s}^{-1}$$

if P_a is in Torr and D, L in cm, and

$$C_{vltN_2} = 1.41 \times 10^3 \frac{D^4 P_a}{L} \text{ m}^3 \cdot \text{s}^{-1} \tag{2.26}$$

if P_a is in Pa and D, L in m.

In case of a short tube, the conductance C_{vstN_2} in the viscous flow region is given by

$$C_{vstN_2} = \frac{\pi}{128} \frac{D^4 P_a}{\eta L} \left[1 + 1.14 \frac{m}{8\pi\eta kT} \frac{Q}{L} \right]^{-1} \quad (\text{in cgs units}) \tag{2.27}$$

Equation (2.27) is valid for $L > 0.304\, mQ/\pi\eta kT$.

2.2.3 Conductance in the Transition Flow Region

Conductance of a Long Tube

Knudsen's semi-empirical equation in the transition flow range for the conductance C_{tlt} of a long tube is given by

$$C_{tlt} = C_{vlt} + ZC_{mlt} \tag{2.28}$$

where C_{vlt} is the viscous conductance of the long tube; C_{mlt} is the molecular conductance of the long tube and

$$Z = \frac{1 + \left(\frac{DP_a}{\eta}\right) \times \left(\frac{m}{kT}\right)^{\frac{1}{2}}}{1 + 1.24 \times \left(\frac{DP_a}{\eta}\right) \times \left(\frac{m}{kT}\right)^{\frac{1}{2}}} \quad \text{(in cgs units)} \tag{2.29}$$

where m is the mass of the gas molecule; k is the Boltzmann constant; T is the temperature of the gas in Kelvin; D is the diameter of the tube; P_a is the average pressure of the gas; η is the coefficient of viscosity of the gas.

$$\text{Let } Y = \left(\frac{DP_a}{\eta}\right) \times \left(\frac{m}{kT}\right)^{\frac{1}{2}}$$

then

$$Z = \frac{(1+Y)}{(1+1.24Y)}$$

From equations (1.2), (1.7) and (1.15), we have

$$\eta = \lambda' P_a \left(\frac{2m}{\pi kT}\right)^{\frac{1}{2}} \quad \text{(in cgs units)} \tag{2.30}$$

where λ' is the average mean free path at P_a

$$\text{As } Y = \left(\frac{DP_a}{\eta}\right) \times \left(\frac{m}{kT}\right)^{\frac{1}{2}}$$

$$= \frac{DP_a}{\lambda' P_a \left(\frac{2m}{\pi kT}\right)^{\frac{1}{2}}} \times \left(\frac{m}{kT}\right)^{\frac{1}{2}}$$

$$= 1.25 \frac{D}{\lambda'}$$

$$\text{Thus} \qquad Z = \frac{\left(1+1.25\dfrac{D}{\lambda'}\right)}{\left(1+1.55\dfrac{D}{\lambda'}\right)}$$

Substituting the value of C_{vlt} from equation (2.25), and the value of C_{mlt} from equation (2.17), in equation (2.28), we have

$$C_{tlt} = \left[\frac{\pi D^4 P_a}{128\eta L}\right] + \left[\frac{\left(\dfrac{2\pi kT}{m}\right)^{\frac{1}{2}} D^3}{6L}\right]\left[\frac{\left(1+1.25\dfrac{D}{\lambda'}\right)}{\left(1+1.55\dfrac{D}{\lambda'}\right)}\right] \qquad (2.31)$$

(in cgs units)

This equation is valid for all types of gas flow.

From equations (2.25) and (2.17), we have

$$C_{vlt} = \frac{\pi D^4 P_a}{128\eta L}$$

$$C_{mlt} = \left(\frac{2\pi kT}{m}\right)^{\frac{1}{2}} \frac{D^3}{L}$$

$$\frac{C_{vlt}}{C_{mlt}} = 0.0736\frac{D}{\lambda'} \qquad (2.32)$$

$$\text{As} \quad C_{tlt} = C_{vlt} + ZC_{mlt}$$

using value of C_{vlt} from equation (2.32), we have

$$C_{vlt} = C_{mlt}\left(0.0736\frac{D}{\lambda'}\right) + Z_{Cmlt}$$

$$= C_{mlt}\left(0.0736\frac{D}{\lambda'} + Z\right)$$

$$= C_{mlt}\left[\left(\frac{0.0736D}{\lambda'}\right) + \frac{\left(1+1.25\dfrac{D}{\lambda'}\right)}{\left(1+1.55\dfrac{D}{\lambda'}\right)}\right]$$

Hence

$$C_{tlt} = C_{mlt}\left(0.0736\frac{D}{\lambda'} + 1\right) \tag{2.33}$$

2.3 Basic Pumping Equation

The basic pumping equation is expressed as

$$V\frac{dP}{dt} = -PS + Q \tag{2.34}$$

where V is the volume to be pumped; P is the pressure at time t; S is the pumping speed; Q is the total rate of influx (throughput) of gas.

The influx of gas can be due to leaks, thermal outgassing from the walls, diffusion from the walls, desorption from other sources.

At ultimate pressure P_u, $dP/dt = 0$. Hence $P_uS = Q$ and

$$P_u = \frac{Q_g}{S} \tag{2.35}$$

If there are no leaks, the influx rate Q_g is due to outgassing from walls, diffusion from the walls and desorption from other sources. Also, at equilibrium pressure $P_e > P_u$, in the presence of the leak (intentional or otherwise),

$$P_e = \frac{Q}{S} \tag{2.36}$$

2.4 Standard Leaks

Standard leaks with known leak rates are used for the calibration of leak detectors and determination of conductance. These employ the permeation of helium through glass or Teflon. Also employed are metal capillaries and micro-tube capillaries of acrylic, polystyrene and polycarbonate. Standard leaks are commercially available in the range 10^{-8}–10 Pa · liters · s^{-1}. Permeation-based standard leaks are sensitive to small temperature fluctuations and can be used only with those gases for which permeating materials are available. Physical leaks are susceptible to clog when exposed to air at atmospheric pressure[1]. Nano-holes[2], with diameters less than or equal to 200 nm, can be utilized in the molecular flow regime up to atmospheric pressure and do not clog. The nano-holes are manufactured by milling a silicon nitride membrane by means of a focused ion beam, and their shapes are characterized by both scanning electron microscopy and atomic force microscopy[3].

The leak element is generally connected to a small reservoir filled with high-purity helium or other gas at a known pressure P_0 and the other end of the standard leak assembly with conductance C_L is connected to vacuum. As P_0 is higher by a few orders of magnitude than the pressure in the vacuum system, the leak rate Q_L of the standard leak is given by

$$Q_L = P_0 \times C_L$$

2.5 Rise of Pressure in a Sealed-off Vacuum Device

In a sealed-off vacuum device such as a vacuum interrupter or an X-ray tube, the base pressure in the device is considered to be in equilibrium with the rate of influx of gas from the interior of the device and the residual pumping speed that is offered by the gettering action in the device assembly, assuming that there is no leak of air from the exterior of the vacuum device.

If q_g is the specific outgassing rate in Pa·m·s^{-1} from the internal surfaces of area A of walls of the vacuum chamber, then Q_g, the total outgassing rate is given by

$$Q_g = q_g \times A \quad \text{in Pa} \cdot \text{m}^3 \cdot \text{s}^{-1}$$

The base equilibrium pressure P_e will be given by

$$P_e = \frac{Q_g}{S_r} \quad \text{in Pa}$$

where S_r is the residual pumping speed in $\text{m}^3\text{·s}^{-1}$ offered by the gettering action.

The pressure will rise above P_e if the atmospheric air leaks into the device. If Q_L is the rate of influx due to the leak, the pressure will rise in the device based on the relation below.

$$V\frac{dP}{dt} = -PS + Q_L$$

where P is the pressure of the gas inside the device at time t, V is the volume of the device. Assuming that the pumping speed S_r is zero or very small and $Q_L \gg Q_g$, this relation can be expressed as

$$V\frac{dP}{dt}=Q_L$$

$$Q_L = C_L\left(P_A - P\right)$$

where C_L is the leak conductance and P_A is the atmospheric pressure. As $P \ll P_A$,

$$Q_L = C_L \times P_A$$

Thus

$$V\frac{dP}{dt}=Q_L$$

and

$$\frac{dP}{dt}=\frac{Q_L}{V}$$

Since the terms Q_L and V are constant, the pressure P in the device will rise linearly with time above P_e as long as $P \ll P_A$. Thus pressure P in the device at any time t after introduction of the leak can be calculated. Also, the time period required to attain a certain pressure can be calculated if Q_L is known.

Consider a sealed-off vacuum device of volume 1×10^{-3} m^3 at room temperature with a residual pressure of 1×10^{-4} Pa. Let us calculate the time that will be required to attain a pressure of 1×10^{-1} Pa in the device with the onset of a leak Q_L of magnitude 1×10^{-11} Pa · m^3 · s^{-1}.

The amount of gas present in the device before onset of the leak is 1×10^{-7} Pa · m^3. The amount of gas in the device on attaining the pressure of 1×10^{-1} Pa will be 1×10^{-4} Pa · m^3. The amount of gas dPV introduced into the device due to the leak is $(10^{-4}–10^{-7})$ Pa · m$^3 \approx 10^{-4}$ Pa · m^3.

The time t required to attain the pressure of 1×10^{-1} Pa is given by

$$t=\frac{dPV}{Q_L}=\frac{10^{-4}}{10^{-11}}\text{s} = 1\times 10^7\,\text{s} = 115.7 \text{ days}$$

References

1. N. Wilson and L. Beavis, Handbook of Vacuum Leak Detection, American Vacuum Society **55** (1976).
2. V. Ierardi, G. Firpo and U. Valbusa, J. Phys.: Conf. Ser. **439**, 012033, (2013).
3. G. Binnig, C. F. Quate, and Ch Gerber, Physical Review Letters **56**, 930, (1986).

3

Surface Phenomena

3.1 Neutrals–Surface Interactions

3.1.1 *Diffusion of Gases*

Diffusion of gases occurs from the region of higher concentration of the gases to the region of lower concentration. Fick's first law of diffusion states

$$q = -D\frac{dc}{dx} \tag{3.1}$$

where q is the gas flow across an area of unit cross section in unit time and D is the diffusion constant with dimensions L^2T^{-1}. Henry's law states that

$$c = sP^n \tag{3.2}$$

where c stands for concentration, P for gas pressure and s for solubility. For non-metals, $n = 1$. Disassociation of diatomic molecules occurs in metals and the concentration is proportional to the square root of the pressure. Thus $n = 1/2$ with units as $atm^{1/2}$.

Concentration c is the amount of gas dissolved in the unit volume of the solid at 293 K. Solubility s is the volume of gas dissolved per unit volume at STP (293 K and 1 atm) of the solid.

The diffusion constant D changes exponentially with temperature and is given by

$$D = D_0 \exp\left(\frac{-E}{RT}\right) \tag{3.3}$$

where D_0 is the proportionality constant, E is the activation energy for diffusion, R is the gas constant and T is the temperature in Kelvin.

3.1.2 Permeation

Permeation is a process in which the gas incident on the solid surface of a wall emerges from the other surfaces of the wall after entering and traversing through the thickness of the wall.

Consider a slab of a large surface area and thickness d. P_1 and P_2 are the gas pressures on two faces of the wall, with surface concentrations of c_1 and c_2. Using Henry's law, we have

$$c_1 = sP_1^n \quad \text{and} \quad c_2 = sP_2^n$$

From equation (3.1), we have

$$q\int_0^d dx = -D\int_{c_1}^{c_2} dc \tag{3.4}$$

Thus, considering an area of unit cross section, we have

$$q = \frac{D_p\left(P_1^n - P_2^n\right)}{d} \tag{3.5}$$

n =1 for non-metals and n = 0.5 for diatomic molecules in metals. D_p is the permeation constant referred to as K and is expressed as the amount of gas in cm^3 at STP diffusing through 1 cm^2 cross section of a slab, 1 cm thick for a pressure difference of 1 atm across the slab. Thus,

$$q = \frac{K\left(P_1^n - P_2^n\right)}{d} \tag{3.6}$$

where K is defined as the permeation constant. Permeation constants as a function of temperature for the gas–non-metal combination and the gas–metal combination have been reported[1,2,3,4]. Also, the solubilities for various gas–non-metal and gas–metal combinations as a function of temperature have been reported[1].

Significant data[5] is available on the permeation of air through various vacuum sealing gasket materials. Altemose[2] has determined helium permeation rates through 20 different glasses using a mass spectrometer. It is interesting to note that the ultimate vacuum produced in the glass vacuum system is limited by permeation of the atmospheric helium through glass[6,7]. It is also observed that there is a limited permeation of H_2 through high-alumina ceramics[8,9]. A review on the subject of permeation is given by Norton[10].

3.1.3 Physical and Chemical Adsorption

Adsorption is a process in which the gas molecules impinging on the solid surface are held on the surface by forces of attraction normal to the surface. Figure 3.1 shows how a potential energy of a molecule varies with its distance from the solid surface. The incoming molecule is attracted towards the surface and remains stable at the minimum potential energy position. The energy H_A is the heat of physical adsorption, generally expressed in kilocalories per mole. This energy equals the energy E_{DP} of desorption.

The process of physical adsorption involves values of heat of adsorption, up to a few kcal/mole. Physical adsorption involves van der Waals intermolecular forces. Multilayer adsorption can occur with still lower values of H_A. Lennard-Jones[11] has regarded the metal surface as a polarized body and has considered the interaction between the incident molecule and the metal surface using electrical image forces. Bardeen[12] has treated the adsorbed atom–metal system on the quantum mechanical basis. Margenau and Polland[13] divided the metal surface into small elements and have considered the interaction

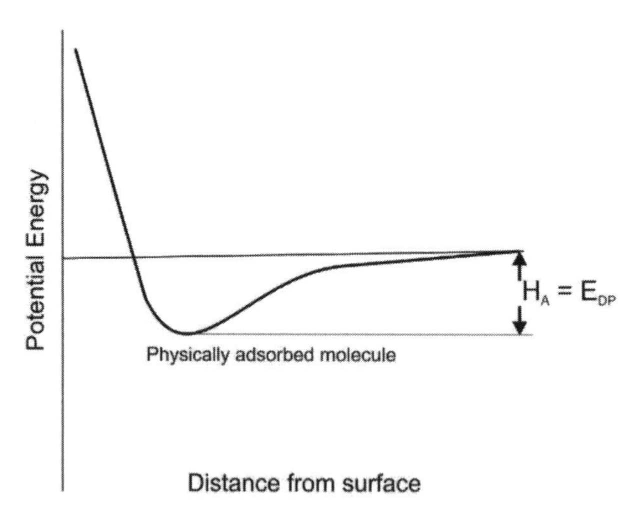

Fig. 3.1. Variation of the potential energy of molecule with distance from solid surface – physical adsorption.

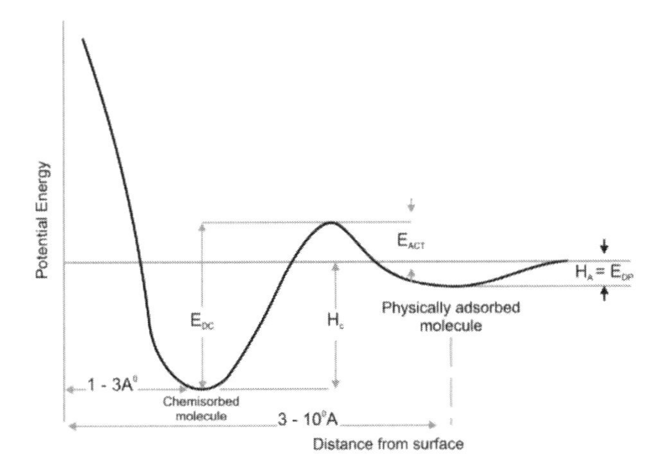

Fig. 3.2. Variation of the potential energy of the molecule with distance from the solid surface – activated chemisorption.

between the adsorbed atom and the instantaneous dipole fields individually in each element.

The process of chemical adsorption (chemisorption) is similar to chemical compound formation and involves relatively higher values of heat of adsorption, up to a few hundreds of kcal/mole. The process initiates with physical adsorption of the molecule at the first potential energy minimum as discussed. Further, on supplying activation energy E_{ACT}, chemisorption takes place as the molecule stabilizes at the next minimum potential energy position as shown in Fig. 3.2. In this case of activated chemisorption, H_C is the heat of chemisorption. The energy of desorption E_{DC} equals the sum of E_{ACT} and H_C. Chemisorption[14,15] is a strong (chemical) interaction between the adsorbed molecule and the surface molecule of the active site. It is selective and occurs only on certain clean surfaces. It is a single layer process, often requiring an activation energy. It can take place at elevated temperatures and is always localized. The chemisorption sites may be homogeneous or heterogeneous according to the energy characteristics of the adsorption site.

A hydrogen molecule initially approaches the tungsten surface along the physisorption curve. It may fall straight into the chemisorption well if it has sufficient energy. Alternatively, it may first experience transient physisorption and it can then either desorb back as a molecule into the gas phase or surpass the barrier to occupy the further lower potential energy minimum into a dissociated state involving chemisorption. Hydrogen is adsorbed on the tungsten surface in the atomic form. Collision of two adsorbed atoms on the surface may cause their recombination to form a molecule and can desorb as a molecule[16]. Ehrlich[17] has investigated the activation energy for surface migration, the energy of desorption of Ca, W, Ba, O, H, N, CO, Xe, Kr, A on W and the heats of adsorption of Rb, Cs, B, H_2, O_2, CO, N_2, CO_2, NH_2, Xe, Kr, A on W.

3.1.3.1 Monolayer

When the surface of a solid is covered by a single layer of adsorbed gas, it is called a monolayer. The nitrogen molecule has its kinetic diameter 3.7×10^{-8} cm. Assuming this equals the distance between the centres of two adjacent molecules in a row or column of the surface, the monolayer will contain 7.3×10^{14} molecules per square centimeter. This value will depend upon the lattice constant of the adsorbent and the molecular size and the average value is taken as 5×10^{14} molecules · cm^{-2} or 5×10^{18} molecules · m^{-2}. The monolayer formation time for gases on a clean solid surface depends upon the gas, its temperature and pressure.

3.1.3.2 Sticking probability

The sticking probability s_t is defined as the probability of adsorption of an incident molecule on the surface. Measurement of s_t is made by the measurement of sorption/pumping speed S of the surface and comparing it with conductance C_0 of an orifice of the equivalent size. Thus

$$s_t = \frac{S}{C_0}$$

Several workers[18,19] have investigated the subject of sticking probability. Generally, the sticking probabilities vary between 0.1 and 1 and then fall as the monolayer coverage is approached.

We have seen that number of molecules ν incident on unit area per second is given as follows by equation (1.9)

$$\nu = P(2\pi mkT)^{-\frac{1}{2}}$$

If s_t is the sticking probability of the molecules on the surface, the rate of adsorption on the unit area will be

$$\frac{dn_a}{dt} = s_t P(2\pi mkT)^{-\frac{1}{2}} \tag{3.7}$$

where P is the pressure of the gas; m is the mass of a molecule; k is the Boltzmann constant; T is the temperature of gas in Kelvin

Using equation (1.10) we have the rate of adsorption $\frac{dn_a}{dt}$ as

$$\frac{dn_a}{dt} = s_t \times 2.63 \times 10^{24} P(MT)^{-\frac{1}{2}} \quad \text{m}^{-2} \cdot \text{s}^{-1} \quad \text{if } P \text{ is in Pa} \tag{3.8}$$

where M is the molecular weight of the gas.

For nitrogen at a temperature of 293 K, and assuming the unit sticking probability the rate of adsorption at a pressure of 1.33×10^{-8} Pa will be

$$\frac{dn_a}{dt} = 3.86 \times 10^{14} \text{m}^{-2} \cdot \text{s}^{-1}$$

At this rate, the monolayer for nitrogen (7.3×10^{18} molecules $\cdot$ m^{-2}) will be formed at a temperature of 293 K in time

$$t = 1.89 \times 10^4 \text{s} = 315 \text{ minutes (5 hours, 15 minutes)}$$

This indicates that a clean surface can remain clean for a relatively longer period of time at lower pressures for a given gas temperature. The adsorbed molecules reside on the surface for a finite period of time depending on the temperature T_s of the surface. The average residence time τ_a of a molecule is given by Frenkel[20] as

$$\tau_a = \tau_0 \exp\left(\frac{E_d}{RT_s}\right) \tag{3.9}$$

where E_d is the energy of desorption, T_s is temperature of the surface and τ_0 is the period of oscillation of the molecule, normal to the surface, typically in the range of 10^{-12} to 10^{-13} s.

Surface coverage fraction θ of the surface is defined as θ = Number of adsorption sites occupied/number of sites available.

Also, θ can be considered as product: the molecular impingement rate $\times$ residence time

From equations (1.9) and (3.9), we have

$$\theta = \nu\tau_a \tag{3.10}$$

$$= \tau_0 \exp\left(\frac{Ed}{RT_s}\right) P(2\pi mkT)^{-\frac{1}{2}} \tag{3.11}$$

This assumes the unit sticking probability for all the molecules that hit the surface.

3.1.3.3 Adsorption Isotherms and Surface Area

Generally, isotherms illustrate the variation of the amount of gas adsorbed on surface or the surface coverage θ with the pressure of gas at selected temperatures. The linear adsorption isotherm[21] is the simplest adsorption isotherm in that the amount of the gas adsorbed on the surface is proportional to the pressure of the gas.

The linear adsorption isotherm is expressed by the equation:

$$\theta = K_H P \tag{3.12}$$

where θ is the surface coverage fraction; P is the pressure of gas; K_H is Henry's adsorption constant.

The linear adsorption isotherm can be considered as the initial stage of other adsorption isotherms and is valid for low surface coverage. Figure 3.3 illustrates the linear adsorption isotherm.

The surface coverage increases with pressure till saturation is achieved when a monolayer is formed. Such isotherm is of the Langmuir type[22], as shown in Fig. 3.4 and is expressed by the equation

$$\theta = bP(1 + bP)^{-1} \tag{3.13}$$

where P is the pressure of the gas and b is a constant.

The Brunauer, Emmet, and Teller[23] (BET) isotherm is a general multi-layer model. It assumes that the Langmuir isotherm model is applicable to each layer and considers vertical atomic interactions while ignoring the horizontal ones. The BET equation is

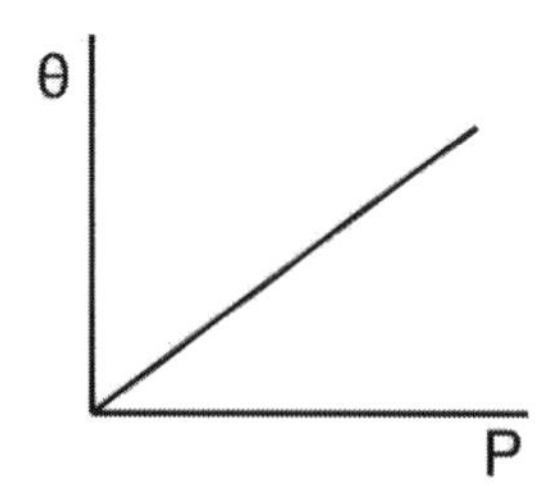

Fig. 3.3. Linear adsorption isotherm.

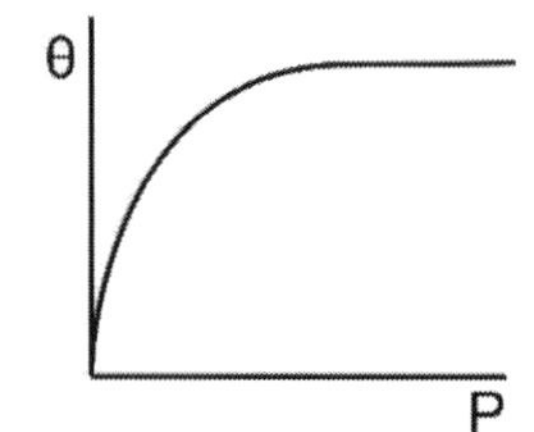

Fig. 3.4. Langmuir type adsorption isotherm.

$$\frac{1}{v\left[\frac{P_0}{P_1}-1\right]}=\left(\frac{P}{P_0}\right)\frac{(c-1)}{v_m c}+\frac{1}{v_m c} \tag{3.14}$$

where P and P_0 are the equilibrium and the saturation pressures of the gas at the temperature of adsorption, v is the adsorbed gas quantity (in volume units), v_m is the monolayer adsorbed gas quantity and c is the BET constant. Figure 3.5 shows the BET isotherm.

The Freundlich adsorption isotherm[24] is mathematically expressed as

$$\theta = KP^{\frac{1}{n}} \tag{3.15}$$

where K is a constant related to the adsorption capacity, P is the gas pressure and n is a constant, generally greater than unity.

The Freundlich adsorption isotherm is illustrated in Fig. 3.6. Values of the heat of adsorption, the monolayer capacity, the residence time of molecules on a surface can be computed from BET isotherm. The actual surface area of a solid surface is much larger than the geometrical surface area. The solid surfaces are not flat on the atomic level. The surface area[25] of irregular surfaces or powders can be determined using models such as the BET (Brunauer-Emmett–Teller) isotherm.

Steininger et al[26] have investigated the adsorption system Pt(111)-CO with LEED, thermal desorption and vibration spectroscopy using electron energy loss spectroscopy (EELS). They observed spontaneously ordered overlayers of CO upon adsorption at about 100 K at low coverages ($\theta \lesssim 0.33$). They further reported that at higher coverages the superstructures exhibited a substantial disorder at low temperatures, but transformed into well-ordered structures with defined diffraction patterns upon annealing and

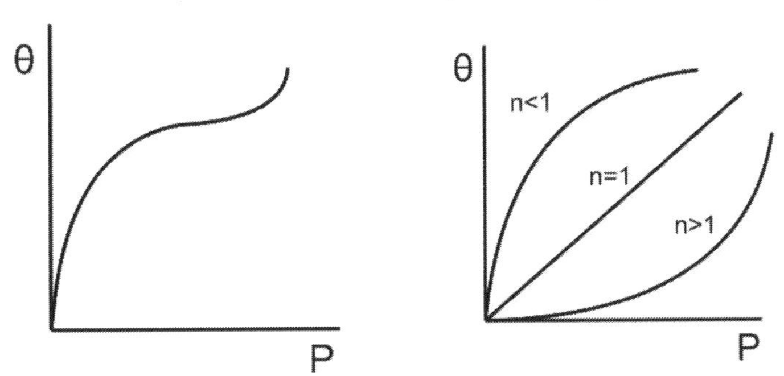

Fig. 3.5: BET adsorption isotherm. **Fig. 3.6:** Freundlich adsorption isotherm.

subsequently cooling the sample. The structures were characterized according to the binding types of CO by their vibrational spectra. At 300 K, they observed that the sticking probability is almost coverage independent within a certain range. King and Well[27] have investigated the adsorption kinetics of nitrogen on tungsten using the molecular beam in UHV. For nitrogen on a polycrystalline tungsten foil, sticking probabilities and their coverage profiles were found to be independent of the incident molecular beam angle, over a range of 60°, with an initial sticking probability s_0 at 300 K of 0.61 ± 0.02. On the W(111) plane s_0 was found to be 0.08 ± 0.01, and further declined linearly with increasing coverage, with a saturation coverage of 1.2 (± 0.2) × 10^{14} molecules · cm^{-2}. They explained that adsorption sites on the (111) plane are created by the presence of surface vacancies or randomly distributed W atoms on the perfect (111) surface.

3.1.4 Desorption

Desorption is the process by which the adsorbed gas is released by providing energy of desorption E_d. If the energy is provided by external heat, the process is termed as thermal desorption. Other desorption processes include energetic gas particle-induced, electron-induced, ion-induced and radiation-induced desorption.

Consider a situation having a surface with fractional coverage ($\theta < 1$) and desorption without recombination. Assuming a first order desorption, we have the desorption rate dn_d/dt given by

$$\frac{dn_d}{dt} = \frac{N_0\theta}{\tau_a} \tag{3.16}$$

where N_0 is the number of molecules in one monolayer on the surface per unit area = 5 × 10^{14} cm^{-2} or 5 × 10^{18} m^{-2} (the amount of gas for 5 × 10^{18} m^{-2} is 2 × 10^{-2} Pa · m); θ is the surface coverage fraction, and τ_a is the average sojourn time.

At room temperature, we have

$$\frac{d(PV)_d}{dt} = \frac{2\times10^{-2}\,\theta}{\tau_a} \quad \text{Pa}\cdot\text{m}\cdot\text{s}^{-1} \tag{3.17}$$

Frenkel[20] has shown that

$$\tau_a = \tau_0 \exp\left(\frac{E_d}{RT}\right) \tag{3.18}$$

where T_s is temperature of the surface and τ_0 is the period of oscillation of the molecule normal to the surface, about 10^{-13} s.

Thus, we have

$$\frac{dn_d}{dt} = N_\theta \tau_0^{-1} \exp\left(\frac{-E_d}{RT_s}\right) \tag{3.19}$$

Equation (3.19) is applicable to the fractional surface coverage of the physically adsorbed and first order chemisorbed gas. For the second order molecular desorption of the adsorbed atoms, we have

$$\frac{dn_d}{dt} = \gamma (N_0\theta)^2 \exp\left(\frac{-E_d}{RT_s}\right) \tag{3.20}$$

where γ is the rate constant.

Gassing of an idealized system resulting from a first order desorption of a monolayer on the wall was calculated by Hobson[28] assuming: volume: 1 liter; surface area: 100 cm^2; sticking probability: 0.5; T: 295 K; pumping speed: 1 liter·s^{-1}

It was established that

- Gas is removed at a much faster rate for $E_d < 15$ kcal/mole.
- The removal rate of the gas is slower for the value of $E_d = 17$ kcal/mole and still further slower for $E_d = 18$ kcal/mol.
- Gas for $E_d > 18$ kcal/mole continues to be released at a slower rate.
- The gassing is negligibly small for $E_d > 25$ kcal/mole.

Thus the factor $E_d/T_s < 0.05$ applies for a faster pumpdown, or $E_d/T_s > 0.083$ applies for negligible gassing.

This indicates that a slowly desorbing gas with $E_d = 30$ kcal/mole can be as easily desorbed as a gas with $E_d = 15$ kcal/mole, at 295 K if the temperature is raised. A few hours bakeout of vacuum systems at 300°C to 450°C is much effective in reducing the gassing rates by a few orders of magnitude, thereby facilitating the achievement of lower pressures.

Let us consider an equilibrium condition between adsorption and desorption for the non-activated first order adsorption of fractional coverage. Equating relations (3.8) and (3.19), we have

$$2.63 \times 10^{24} s_t P(MT)^{-\frac{1}{2}} = N_0 \theta \tau_0^{-1} \exp\left(\frac{-E_d}{RT_s}\right)$$

The number of molecules on a surface of 1 m^2 is given by

$$N_0\theta = 2.63\times10^{24}\, s_t\, P(MT)^{-\frac{1}{2}}\, \tau_0 \exp\left(\frac{E_d}{RT_s}\right) \tag{3.21}$$

Or

$$\theta = \frac{2.63\times10^{24}\, s_t\, P(MT)^{-\frac{1}{2}}\,\tau_0}{N_0}\exp\left(\frac{E_d}{RT_s}\right) \tag{3.22}$$

As $N_0 = 5 \times 10^{18}$ molecules · m^{-2},

$$\theta = 5.26\times10^{5}\, s_t\, P(MT)^{-\frac{1}{2}}\,\tau_0 \exp\left(\frac{E_d}{RT_s}\right) \tag{3.23}$$

where T is the gas temperature, T_s is the surface temperature and P is the pressure in Pa.

The amount of gas $N_0\theta$ adsorbed can be calculated by using equation (3.21) as a function of temperature or pressure by plotting the coverage $N_0\theta$ as a function of pressure at constant temperature and as a function of temperature at constant pressure. Also, the pressure can be plotted as a function of temperature at constant coverage $N_0\theta$.

The upper limit for the gas that can be physically adsorbed at room temperature can be determined by equation (3.23). Sojourn time τ_a for a given E_d and temperature T_s of the surface can be calculated. Further, θ can be calculated for the gas with molecular weight M at temperature T by assuming $s = 1$.

For effective adsorption, the rate of adsorption must exceed the rate of desorption:

$$\frac{dn_a}{dt} \gg \frac{dn_d}{dt}$$

From equations (3.8) and (3.19), this can be expressed as

$$2.63\times10^{24}\, s_t\, P(MT)^{\frac{-1}{2}} \gg N_0\theta\,\tau_0^{-1} \exp\left(\frac{-E_d}{RT}\right) \tag{3.24}$$

Substituting the value of τ_0 exp (E_d/RT) from equation (3.18), we have

$$2.63\times10^{24}\, s_t\, P(MT)^{-\frac{1}{2}} \gg N_0\theta\,\tau_a^{-1} \tag{3.25}$$

$$\tau_a \gg \frac{N_0\theta}{\left[2.63\times10^{24}\, s_t\, P(MT)^{-\frac{1}{2}}\right]}$$

$$\tau_a \gg \frac{\left[5\times10^{18}\theta(s_tP)^{-1}(MT)^{\frac{1}{2}}\right]}{2.63\times10^{24}}$$

$$\tau_a \gg 1.9\times10^{-6}\theta(s_tP)^{-1}(MT)^{\frac{1}{2}} \qquad (3.26)$$

Consider; $s_t = 0.5$; $P = 10^{-3}$ Pa; $M = 28$; $T = 300$ K. Thus $\tau_a \gg 0.35$ s.

It can be seen by using relation (3.18) that at the surface temperature of 300 K and the sojourn time of 0.35 s, the desorption energy corresponds to about 17 kcal/mole (0.74 eV/atom). Thus for effective adsorption under these conditions, it is necessary to have $\tau_a \gg 0.35$ s or $E_d \gg 17$ kcal/mole (0.74 eV/atom).

3.1.4.1 Determination of Activated Energy of Desorption

Methods for determining the activation energy, the rate constant and the order of reaction from thermal desorption experiments are examined by Redhead[29] by using two heating schedules. The activation energy of desorption is estimated from the temperature (T_p) at which the desorption rate is a maximum (for first-order reactions) and from the change of T_p with surface coverage (for second-order reactions). The order of the reaction is determined from the shape of the desorption rate versus time curve, or from the variation of T_p with coverage.

The temperatures T_p at which peaks occur in the desorption spectra are determined experimentally. The values of the activation energy E_d for desorption are calculated at different values of T_p by using the relation given by Redhead[29]

$$\frac{E_d}{kT_p} = \left(\frac{YT_P}{\beta}\right) - 3.64 \qquad (3.27)$$

where k is the Boltzmann constant (8.6×10^{-5} eV/K), γ is the first order rate constant in s^{-1} and β is the rate of rise of the temperature of the target in $K \cdot s^{-1}$. E_d and T_p are expressed in eV/atom and K respectively. Relation (3.27) is derived from the relation of first order desorption from the unit surface area of the solid.

$$\frac{d\sigma}{dt} = -\sigma Y \exp\left(\frac{-E_d}{kT_p}\right) \qquad (3.28)$$

where σ denotes the number of atoms trapped per cm^2 and T is the temperature of the solid in Kelvin.

Relation (3.28) holds good for the desorption of atoms trapped within few tens of lattice constants from the surface of the solid.

Redhead[29] has given details of the numerical solution for the case

$$T = T_0 + \beta t$$

where T_0 is the starting temperature of the heating cycle in Kelvins, t being the time period in seconds at temperature T.

Relation (3.28) is solved to find the temperature T_p at which the desorption rate is maximum. Redhead[29] has further shown that the relation between E_d and T_p is very linear and is expressed by the relation (3.27) for

$$10^{13} > \frac{Y}{\beta} > 10^{8}\ K^{-1} \text{ to } \pm 1.5\%$$

Values of E_d are determined from the desorption spectra assuming $\gamma = 10^{13}$ s^{-1} and $\beta = 42.5$ K · s^{-1}.

Cowin et al[30] have applied the laser heating technique for studying fast surface processes in an initial study to the thermal desorption of D_2 from a polycrystalline tungsten sample. The surface is subjected to a sufficiently fast and a large temperature rise to desorb surface atoms or molecules in a time short compared to the range of flight times to a mass spectrometer detector. In this way the velocity distribution of the desorbing species can be determined. They used a laser pulse width of 3×10^{-8} s and a surface temperature rise of 300 to 3000 K. George et al[31] have used a pulsed laser to desorb hydrogen from a small region on the surface of Ni(100). Subsequent laser induced desorption measurements led to the determination of the time required to refill the vacant area by hydrogen from nearby regions and measurement of the surface diffusion coefficient for hydrogen at temperatures between 223 K and 283 K.

3.1.4.2 Photon–Electron and Ion-Induced Desorption

Photons can cause desorption as a result of a quantum interaction, as opposed to thermal effects. For a sufficiently high energy, the adsorbate can be excited and can form an excited atom or neutral atom or ion, followed by desorption.

Two different modes of electronic transition have been studied to explain electron-induced desorption (EID). These are

- The core–electron excitation and
- The valence–electron excitation

The core–electron excitation is followed by an Auger transition that results in the formation of a doubly-ionized anion, with two holes localized on an anion. Knotek and Feibelman[32] have explained that the hole localization leads to desorption due to the Coulomb repulsion between the doubly-ionized anion and neighboring locations. Menzel and Gomer[33] and Redhead[34] have suggested that electronic excitation yields an anti-bonding potential surface that leads to the ejection of an atom from the surface.

EID is fairly independent of the substrate material. Both neutrals and ions can cause desorption. Ion-induced desorption is observed in photon storage rings. Photon-induced desorption occurs in electron storage rings.

Bender et al[35] have shown that in heavy-ion synchrotrons, high-energy ions can impact on the beam pipe and release gas molecules. The ion-induced desorption deteriorates the accelerator vacuum, thereby adversely affecting the beam life time and luminosity. Mahner[36] has investigated molecular desorption induced by heavy ions in particle accelerators.

Desorption of CO and NO molecules chemisorbed on Pt (001) and Pt (111) surfaces at 80 K induced by ultraviolet nanosecond-pulsed laser irradiation has been studied[37] by a resonance-enhanced multi-photon ionization technique. It was observed that in this case the desorption is not thermally driven, but is induced by electronic excitation. On the basis of these results, the desorption mechanism induced by valence–electron excitation for molecules chemisorbed on metals has been discussed in connection with the unoccupied electronic structures of the adsorbate and the solid metal.

3.1.5 Thin Films Deposition on Surfaces

The thin film is a layer of material ranging from a nanometer to several micrometers in thickness. Thin films are deposited on clean substrates in a high-vacuum environment either by the sputtering or evaporation process.

Venable et al[38] and Lewis and Anderson[39] have given a review of the nucleation and growth processes involved in the formation of thin films.

Thin film deposition can be done by thermal evaporation. This can be achieved using a filament as the heating element. Alternatively an electron beam is aimed at the source material to evaporate it and enter the gas phase. The evaporated atoms which are not scattered due to the presence of high vacuum, are condensed onto the substrate where the thin film is formed.

In ion-beam sputter deposition, the ion beam is directed at a target and sputtered atoms are deposited onto a nearby substrate surface. Stuart and Wehner[40] observed that energies of sputtered atoms depend markedly on the angle of ejection and that ejection energies decrease for lighter ions. They also observed that the dependence on bombarding ion energy is very weak above 1000 eV where average energies of sputtered atoms are found to be in the range 5–15 eV. Ejection energies decrease at lower bombarding ion energies. Auciello et al[41] used the ion beam sputter-deposition system to produce W/Cu/W layered films and investigated their microstructure and morphological and electrical characteristics. Layered films with smooth surfaces were deposited on planar surfaces with reasonable conformity. They demonstrated that the W layer microstructure can be tailored by controlling the deposition parameters.

The properties of thin films differ from those of the bulk material. Surface quality, material of the substance, degree of vacuum, flow of reactive gases, rate of deposition, purity of the substance are the important parameters that determine mechanical, optical and other properties.

Kazmerski[42] has discussed in detail the subject of polycrystalline and amorphous thin films and devices. The discussion covers conductors, luminescent thin films, solar energy conversion devices.

Thin films offer wide applications. These include

- protection of surfaces from corrosion, oxidation, wear
- increase in transmission, reflection in certain wavelength regions, filters, colour separation
- high-temperature superconductors, silicon devices, anti-fog, memory devices, nanotechnology, sensors, miniature fuel cells, portable solar cells.

The pulsed laser deposition (PLD) technique uses a plasma 'plume' that propels a mixture of atoms, ions and small particles towards a substrate. S. Alfihed et al[43] have used the PLD technique for investigating the growth of polycrystalline WS_2 films at a relatively low temperature. The PLD chamber is pumped to pressures below 10^{-3} Pa using a turbomolecular pump. The chamber is backfilled with process gases to suit the deposition process. The substrate serves as the seed and the plasma plume serves as the source of the growing crystal that can be doped with rare-earth ions to give the optical activity needed to make a gain medium. The temperature of the substrate is raised to about half of the melting point of the constituent material that is being grown. The plasma plume that hits the target achieves an epitaxial growth on the substrate. The substrate seed is maintained at about 1000°C by using a CO_2 laser. The incoming atoms from the plume move to find their place in the lattice to help build up the final structure.

3.1.6 Molecular Beam Epitaxy (MBE)

Molecular beam epitaxy (MBE) involves the growth of single crystal films and high-purity semiconductor films, on top of a crystalline substrate using deposition of vapours evaporated from Knudsen effusion cells. The deposition is conducted in an ultrahigh vacuum environment to minimize possible contamination from the residual gases to achieve high-purity films. The deposition rate is such that a low growth rate of ~1 monolayer (lattice plane) per second is obtained. The evaporated atoms have long mean free paths and thus they cannot interact with each other or with residual gases in transit till they reach the substrate. A layer by layer growth is achieved. Joyce[44] has reviewed major aspects of MBE.

Single-crystal gallium arsenide can be formed by heating high-purity gallium and arsenic simultaneously in separate Knudsen effusion cells to condense their vapours on the wafer where they react to form the compound[45]. Many modern semiconductor devices are manufactured using the MBE technique. This technique is also used for deposition of oxide materials for electronic, magnetic and optical applications.

3.1.7 Surface Ionization

If neutral gas particles impinge on an incandescent metal surface, they may be released from the surface in the form of neutral atoms, or positive ions or negative ions after a mean residence time. The formation of cesium positive ions on an incandescent tungsten surface was first observed by Langmuir and Kingdon[46]. They observed that the probability of ionization is a function of the temperature, the work function and the ionization energy of the surface.

The Saha–Langmuir equation states that

$$\frac{n_+}{n_0} = \left(\frac{g_+}{g_0}\right) \exp\left[\frac{(W - \Delta E_I)}{kT}\right] \tag{3.29}$$

where n_+/n_0 is the ratio of the ion number density to the neutral number density released per unit area of the surface per second, also called α, the degree of ionization; g_+/g_0 is the ratio of statistical weights (degeneracy) of ionic (g_+) and neutral (g_0); W is the work function of the surface; ΔE_I is the ionization energy of the desorbed gas; k is the Boltzmann's constant; T is the temperature of the surface.

Negative ionization is possible for atoms with a large electron affinity with a surface of low work function.

Also as

$$\alpha = \frac{n_+}{n_0}$$

Further, the ionization coefficient

$$\beta = \frac{n_+}{\nu} \tag{3.30}$$

where ν is the number of atoms impinging the surface per unit area of the surface per second.

As

$$\nu = n + n_0$$

we have

$$\beta = \frac{\alpha}{1+\alpha} \tag{3.31}$$

Thus, β can be expressed[47] as $\beta \approx \alpha$ (as for Ag or Cu on W) if $\alpha \ll 1$ and that $\beta \approx 1$ (as for Cs on W) for $\alpha \gg 1$.

Sroubek[48] has proposed a theory to explain surface ionization and has derived analytical expressions for ionization probabilities of the desorbed particles from surfaces in different conditions such as inactive, excited by independent collisions and thermally excited.

Wendt and Cambel[49] measured potassium ion current densities produced by the surface ionization of potassium on a tungsten filament as a function of the filament temperature, applied potential, and potassium vapour density. These compared qualitatively with the results predicted by the Saha–Langmuir equation and quantitatively with the results predicted by Child's space-charge law.

Ion sources using surface ionization are used in many applications.

3.2 Interaction of Charged Particles, Radiation and Heat with Solid Surfaces

3.2.1 Ion–Surface Interactions

Upon impact with a solid surface, an energetic ion experiences forces of interaction with atoms of the solid, which decide its trajectory.

Elastic and inelastic interactions occur as ions/neutrals bombard surfaces[50]. The following phenomena can occur upon the incidence of energetic (few keV) neutrals/ions on the solid surface.

- Surface ionization
- Charge neutralization of ion and scattering of the incident ion/neutral
- Back-scattering of the incident ion without any loss of charge
- Penetration of the incident ion/energetic neutral into the solid surface. The penetrating particles encounter collisions with the atoms losing energy and causing atomic displacement and defects before their entrapment at some locations in the lattice if these are not backscattered from the surface during their slowing down or if these do not diffuse out from the surface
- Desorption of ions, neutrals
- Sputtering of the solid surface atoms in form of neutrals and ions, positive and negative
- Emission of secondary electrons, X-rays and photons from the solid surface.

The study of ion reflection, penetration, and entrapment offers information on interatomic forces, binding energies, and defect formation. Carter[51] has reviewed this subject.

3.2.1.1 Scattering of Positive Ions

Scattering of positive ions incident on solid surface is studied in different incident ion energy ranges:

- Low-energy ion scattering (LEIS) few hundred eV to few keV
- Medium-energy ion scattering (MEIS) 20–200 keV
- High-energy or Rutherford backscattering (RBS) ~MeV
- Elastic recoil detection analysis (ERD), keV–MeV

If a monoenergetic, singly-charged low-energy (few keV) beam of positive ions of a noble gas with energies in the range 100 eV–10 keV strikes the solid surface in vacuum, some of the incident ions experience binary, elastic collisions with the surface atoms and are back-scattered. The energy spectra of the back-scattered particles at a given scattering angle give direct identification of surface atoms in the first atomic layer.

If E_0 is the energy of the primary ion; E_1 is the energy of the scattered ion; M_0 is the mass of the primary ion; M_s is the mass of the surface atom, and if the scattering angle is 90^0

$$\frac{E_1}{E_0} = \frac{(M_s - M_0)}{(M_s + M_0)}$$

Then

$$M_s = M_0 \frac{\left(1 + \frac{E_1}{E_0}\right)}{\left(1 - \frac{E_1}{E_0}\right)} \tag{3.32}$$

The surface atoms can thus be identified.

In medium-energy ion scattering elastic scattering is considered between two bodies. The energy of the scattered ion is given by

$$E = \left[\frac{\left(m_2^2 - m_1^2 sin^2\theta\right)^{\frac{1}{2}} + m_1 \cos\theta}{m_1 + m_2}\right]^2 E_0 = k^2 E_0 \tag{3.33}$$

where E_0 is the initial ion energy, m_1 and m_2 are the ion and target masses respectively, θ is the scattering angle. The factor k^2 is called the kinematic factor. For two different target masses k^2 varies with increase in the scattering angle. Thus, the scattered ions from two different target masses may be distinguished for sufficiently high scattering angles. Also, for a given target mass, the ion energy decreases with increase in scattering angle.

Rutherford considered the atom as having a very small nucleus containing the positive charge with the electrons orbiting around it. Thus, if the alpha particle comes very close to the nucleus, it would be repelled by a sufficiently large Coulomb force due to the positive nucleus, and would scatter with a large angle. Rutherford estimated the size of the gold nucleus. He developed the Rutherford scattering formula which shows that the curve of the intensity against scattering angle θ is proportional to $1/\sin^4(\theta/2)$.

An energetic (keV–MeV) ion beam causes elastic nuclear interactions with the atoms of the solid. The incident energetic ions possess sufficiently large energy enough to recoil the atoms being struck.

3.2.1.2 Secondary Electron Emission by Ion Surface Interaction

Secondary electrons are generated by neutrals, ions, electrons, or photons with sufficiently high energy. Photoelectrons also can be considered as secondary electrons.

When an ion or an excited atom approaches a metal surface, neutralization of the ion may occur and de-excitation or resonance ionization of the excited atom may take place. It is evident from the experimental data that electronic transitions involved in these processes are almost independent of the kinetic energy of the incident particle and are governed by its potential energy of excitation. The electronic transitions[52,53] include:

- Resonance neutralization
- Resonance ionization
- Auger de-excitation
- Auger neutralization

Secondary electron emission and Auger[54] electron emission follow these processes. The number of secondary electrons emitted per incident particle is called the secondary emission yield. The energy distribution of the secondary electrons can be studied by using a retarding potential that allows the number of electrons having the energy greater than eV to be determined or by using a magnetic field to determine the number of electrons with the energy between eV and eV + deV.

The earliest review on the subject of secondary electron emission has been published by Bruinin[55].

3.2.1.3 *Entrapment of Injected Ions*

Injection of low-energy (up to a few keV) inert gas ions on clean metal surfaces results in the entrapment of the ions at sites near the solid surface. The trapped atoms are released by a programmed heating of the solid. The rate of release of the gas with temperature is monitored by a mass spectrometer to give the desorption spectrum[27]. The temperatures at which the peaks of the rate of gas release occur correspond to the binding energies of particular atoms trapped in the solid with different configurations as discussed in section 3.1.4.1.

Close and Yarwood[56] have reported their results on the entrapment of helium positive ions with the energy of 60 eV–1 keV in polycrystalline tungsten using the thermal desorption method and reported ion trapping probabilities 0.03 at 60 eV and 0.69 at 1 keV. Naik et al[57] have studied the entrapment of inert gas ions of energies between 430 and 1950 eV into polycrystalline molybdenum. They reported that the activation energies of desorption are in agreement with the reported values of the activation energy for migration of atoms and defects in molybdenum. Figure 3.7 shows thermal desorption spectra of Ne^+ ions on polycrystalline Mo from their work. Kornelsen[58,59,60,61] has made extensive experimental investigations of the thermal desorption of ionically trapped gases.

Figure 3.8 shows schematic diagram of thermal desorption system used by Naik et al[56]. Mhaskar and Naik[62] developed a computer program to simulate the activation energies of these detrapping reactions in the bulk of the solid and in the vicinity of the crystal surface. Experimental values of the activation energies of these reactions reported earlier by Kornelsen show good agreement with the values calculated for the same reactions occurring in the bulk. The values calculated for these reactions near the surface are much lower than the experimental values. It is further shown that an agreement between the experimental and the calculated values of activation energies for the surface reactions is possible only if the entrapment is made to occur near the surface. In general, peaks observed in the thermal desorption spectra of helium are related to the detrapping reactions of helium from HenV (n = 1 to 4) complexes. Surface relaxation of W(100) planes and polycrystalline Mo and the effect on point defects were computed by Mhaskar and Naik[63,64] using the

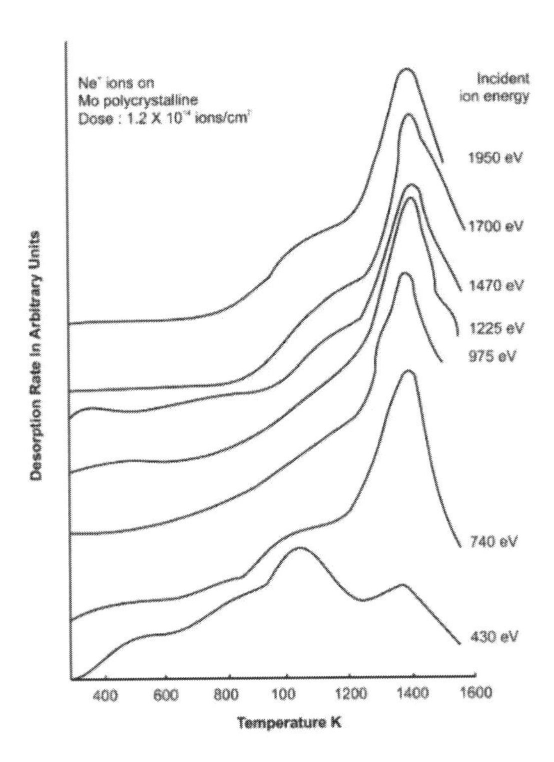

Fig. 3.7. Thermal desorption spectra of Ne+ions on polycrystalline Mo (*Reprinted from Applied Physics December 1980, Volume 23, Issue 4, pp. 373–380 –"Entrapment of Inert Gas ions near molybdenum surface" - P. K. Naik, V. G. Kagal, S. L. Verma, S.P. Mhaskar © by Springer-Verlag 1980 with permission of Springer*).

computer program that can simulate various processes in the bulk of a solid and in the vicinity of the surface.

3.2.1.4 Sputtering

Sputtering involves the emission of particles from the solid surface under the impact of neutrals or charged particles. Typically inert gas ions such as argon ions are used for bombardment of the surface as they are heavier and do no react chemically with the surface atoms. The sputtered material may consist of single, neutral or ionized atoms, and neutral or ionized clusters.

The sputter yield Y is defined as the ratio of the number of emitted atoms to the number of incident particles. The sputter yields are generally in the range of 0.1 to 5.0 for the incident particle energies between 30 eV–1000 eV. These lie between 0.5 and 1.0 for the incident energies between 30 and 1000 eV. For higher incident

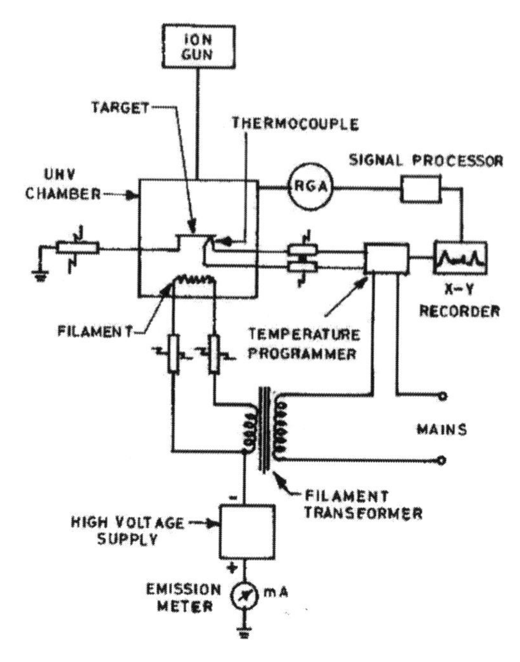

Fig. 3.8: Thermal desorption system (*Reprinted from Applied Physics December 1980, Volume 23, Issue 4, pp. 373–380 "Entrapment of Inert Gas ions near molybdenum surface" – P. K. Naik, V. G. Kagal, S. L. Verma, S.P. Mhaskar © by Springer-Verlag 1980 with permission of Springer*).

particle energies up to 10 keV, the sputter yields[65] vary from about 1.0 to 10.0 depending upon the target material.

In the reactive sputtering process, the atoms sputtered from a metallic cathode react chemically with other gas molecules at the sample surface. Thus the titanium sputter deposited in the presence of nitrogen can form Ti-nitride at the substrate.

Wehner[66] has presented an excellent review of the work in the field up to the year 1955. More recently, Sigmund[67] has reviewed the subject. Sigmund's model of sputtering is based on the linear cascade theory, which yields a linear dependence on the projectile energy. Though the theoretical results were initially verified, in some results later a non-linear dependency of damage[68] as a function of the projectile energy was observed. Merkle and Jäger[69] proposed the presence of high-energy density cascades or sub-cascades, called *surface spikes*, close to the surface. The thermal spike theory explains sputtering in the form of evaporation from a hot surface region. Initially, energy is deposited that generates an initial (vibrational) temperature, after the ion system has reached equilibrium. The hot zone is quenched through heat transport processes, and thermal evaporation takes place on the hot surface region of the spike.

Sputtering phenomena find many applications in science and technology. Secondary ion mass spectrometry (SIMS) is utilized to analyze the composition of solid surfaces and thin films by sputtering the surface of the specimen with a primary ion beam and collecting and analyzing ejected secondary ions. Sputtering is employed in the semiconductor industry to deposit thin films of various materials in integrated circuit processing. Thin anti-reflection coatings and low-emissivity coatings on glass for optical applications are deposited by sputtering. Sputtering is used for depositing the metal layer for fabrication of CDs and DVDs. The sputter-ion pump that utilizes sputtered films of titanium is considered as the work-horse of ultrahigh vacuum technology. The process of sputtering can also be used as a micro- and nano-machining tool, to modify or machine materials at the micro- and nanoscale[70]. It is possible to machine away one atom layer without any disruption of the atoms in the next layer, or any residual disruptions above the surface.

3.2.1.5 Ion Beam Implantation

Ion beam implantation is a process in which accelerated ions are injected into the solid. This method can be employed to cause changes of the physical, chemical, electrical or optical properties of the solid. Ion beam implantation can cause change of crystalline structure of the solid. In 1906, Rutherford conducted the first ion implantation when he bombarded aluminum foil with alpha particles. For penetration of the order of 1 μm, accelerators of 40 kev to 400 keV are required. Such implanters are suitable for metal targets.

Lindhard et al[71] proposed their approach explaining the range of low-energy heavy ions penetrating into solids. It was based on Thomas–Fermi atoms. Later, numerical methods were applied to traditional theoretical approaches[72,73]. The theory was based on statistical models of atom–atom collisions.

Channeling occurs

- when the ion velocity is parallel to a major crystal orientation.
- when some ions may travel considerable distances with little energy loss.

Once in a channel, the ion will continue in that direction, making many glancing internal collisions that are nearly elastic (their stopping is then dominated by the electronic drag only), until it comes to rest or finally dechannels. The latter may be the result of a crystal defect or impurity.

Applications of ion beam implantation include doping of semiconductors and insulators, creation of nanoparticles, change of the electrical and optical properties of solid.

3.2.1.6 Ion Beam Deposition/Ion Plating

This is a process of applying materials to a target using an ion beam. The ion source employs materials in the form of a gas, an evaporated solid, or a solution (liquid). The plasma/gas discharge system can also be used to bombard the target by accelerated ions. The ions are then accelerated, focused or deflected using high voltages or magnetic fields. Deceleration at the target can be employed to control the deposition energy ranging between a few eV up and a few keV. Molecular ion beams are deposited on the surface at low energy.

Aisenburg[74] has discussed mechanisms in the process of ion plating. The ion plating process was first described in the technical literature by Mattox[75]. Ion plating is used to deposit hard coatings of compound materials on tools, adherent metal coatings, optical coatings with high densities, and conformal coatings on complex surfaces.

3.2.2 Electron–surface interactions

Emissions from the solid surface resulting from the incident electrons include

- Back-scattered electrons
- Auger electrons
- Secondary electrons
- Characteristic X-rays
- Bremsstrahlung X-rays
- Cathodoluminiscence (visible light)
- Heat

Some of the incident electrons are diffracted and transmitted through the bulk of the solid. The incident electrons interact with the solid surface atoms and are significantly scattered by them rather than penetrating into bulk of the solid. Those electrons that penetrate the solid surface, lose the energy in heating of the solid by phonon excitation of the atomic lattice. The electrons undergo elastic and inelastic scattering before coming to rest.

In elastic scattering, there are changes in the electron trajectory, however their kinetic energy and the velocity remain constant due to large differences between the masses of the electron and the nucleus. This process is known as electron backscattering. The inelastic interactions give rise to secondary electrons[76] as these occur between the incident electrons and the outer, not strongly bound electrons of the atoms. These outer electrons can be ejected from the atom with energies lower than 50 eV. If these 'secondary' electrons are produced near the surface, and if their energies are higher than the surface energy (~6 eV) then they can escape to vacuum. In case of inelastic scattering, while the trajectory of the incident electron is slightly perturbed, the energy loss is through interactions with the orbital electrons of the atoms in the solid. The elastic and inelastic interactions limit the penetration of the beam into the solid. The region of active interaction between the solid and the beam is known as the interaction volume.

Although secondary and Auger electrons are produced throughout the interaction volume, their energies are too low to escape from the solid surface. Thus such electrons can escape only from a thin layer near the solid surface.

The energy loss through inelastic interactions and electron backscattering through elastic interactions determines the size and shape of the interaction volume. The effective interaction volume is a hemispherical to jug-shaped region with the neck of the jug at the solid surface as illustrated in Fig. 3.9. The regions from where various emissions occur from different depth are shown in the figure. The physical quantities governing the depth ranges indicated in the figure are discussed by Werner[77].

3.2.2.1 Secondary Electron Emission by Electrons

While the transmitted electrons pass completely through the material after interactions, the back-scattered electrons are ejected from the surface of the primary incidence. The energy range of the transmitted and the back-scattered electrons lies between about 50 eV and the accelerating voltage.

The number of secondary electrons emitted per incident particle is called the secondary emission yield. The energy distribution of the secondary electrons can be measured by using a retarding potential

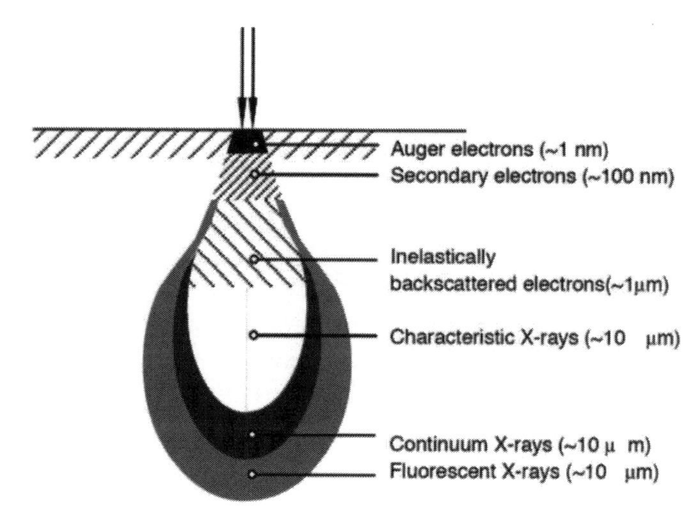

Fig. 3.9. Interaction volume of electrons penetrating the solid surface (*Courtesy: W.S.M. Werner*).

V that allows the number of electrons having the energy greater than eV to be determined or by using a magnetic field to determine the number of electrons with energy between eV and eV + deV.

The earliest review on the subject of secondary electron emission has been published by Bruining[78]. Simon[79] has explained that the secondary-emission process covers physical processes that include the excitation of secondary electrons, transport of these electrons through the solid, and transport through the vacuum–solid interface. He observed that the application of certain solid-state concepts such as band bending has led to enormous improvement in understanding of the transport and escape of secondary electrons and has made possible the development of a new type of secondary emitter with a maximum yield which is an order of magnitude or more greater than the yields of the best previously described materials.

3.2.2.2 Auger Electron Emission

As electrons with energies in the range of several eV to 50 keV are incident on the solid surface, an inner-shell electron is ejected from a sample atom by a primary electron, and an electron from an outer shell fills the vacancy. The energy released from this transition is utilized in emission of an Auger electron or X-ray photon. A second order outer shell electron gains the transition energy and is emitted as an Auger electron from the atom provided the transferred energy

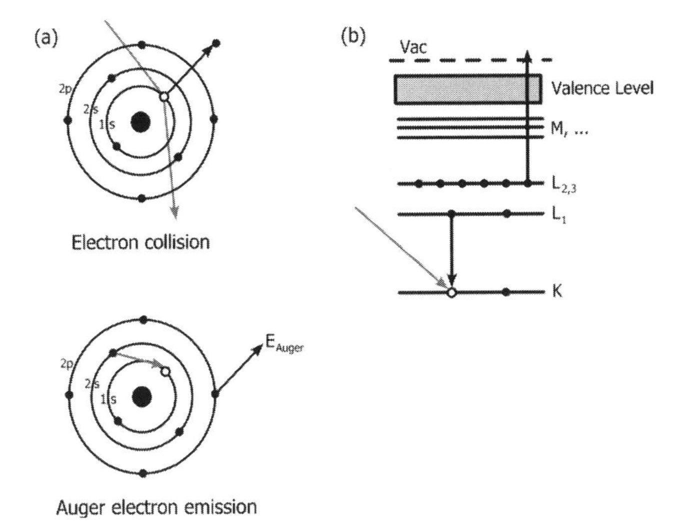

Fig. 3.10. Auger electron emission from solid surface (*Public Domain - A. Carlson (Wikimedia Commons) https://commons.wikimedia.org/wiki/File:Auger_Process.JPG).*

exceeds the orbital binding energy. The emitted electron possesses kinetic energy E_k given by

$$E_k = E_c - (E_b - E_{c'}) \tag{3.34}$$

where E_c is the energy of the core state; E_b is the energy of the first outer shell; $E_{c'}$ is the energy of the second outer shell.

This limits their escape depth. Auger electrons emitted from an interaction in the bulk below the surface lose energy by scattering reactions along their path to the surface. Auger electrons emitted at a depth greater than about 2–3 nm do not possess sufficient energy to escape the surface. The probability is greatest for the emission of an Auger electron for light elements.

The Auger electron emission process is illustrated in Figure 3.10.

3.2.2.3 X-ray Emission

X-rays are produced when electrons of high energy strike a heavy metal target. Some of the electrons approach the nucleus of the metal atoms where they are deflected by the nucleus.

Bremsstrahlung is the electromagnetic radiation produced by the deceleration of a charged particle when deflected by another charged particle, such as an electron, by an atomic nucleus. The moving particle loses kinetic energy, which is converted into a photon.

Bremsstrahlung has a continuous spectrum. Other forms of radiation due to the acceleration of a charged particle include synchrotron radiation, cyclotron radiation, and the emission of electrons and positrons during beta decay. Bremsstrahlung is referred to the radiation from electrons slowing down in matter.

Characteristic X-rays are emitted from heavy elements when their electrons make transitions between the lower atomic energy levels. The characteristic X-rays emission that appear as sharp peaks occur when vacancies are produced in the $n = 1$ or K-shell of the atom and electrons drop down from higher level to fill the gap. The continuous distribution of X-rays which forms the base for the two sharp peaks is called 'bremsstrahlung' radiation.

Most X-rays have frequencies in the range 3×10^{16} Hz to 3×10^{19} Hz and energies in the range 100 eV to 100 keV. X-rays with photon energies above 5–10 keV are called *hard X-rays*, while those with lower energy are called *soft X-rays*. As the wavelengths of the hard X-rays are closer to the size of atoms, they can be used for determining crystal structures by X-ray crystallography.

X-rays interact with matter through photoabsorption, Compton scattering, and Rayleigh scattering. The intensity of the interactions is governed by the energy of the X-rays and the elemental composition of the material. Photoabsorption or photoelectric absorption dominates in the soft X-ray regime and for the lower hard X-ray energies while the Compton effect dominates at higher energies.

3.2.2.4 Interaction of High-Energy Electrons with Surface

Based on the work by Taylor, Slade[80] has shown that high energy (~few hundred keV) electrons can penetrate up to ~180 μm deep in copper. He has further observed that most of the electron beam's energy is deposited to a depth $\sim0.45R_p$ where R_p is the maximum penetration depth of a narrow electron beam.

High-energy (few keV) electrons are used for thermal treatment including welding, annealing, melting[81], machining, etching, thin and thick metal film coating. Naik et al[57] have used electron bombardment using electrons of energy 2 keV for programmed heating for the study of thermal desorption of inert gas ions trapped in molybdenum sample. Reuter et al[82] have used an electron evaporator to generate a molecular beam of carbon.

Electrons with a high power density of $\sim10^7$ W·cm^{-2} are used at the focus of the electron beam for welding applications[83]. With precision electron beam devices, components with dimensions of the order of a 10^{-1} mm can be precisely welded. The energy transfer in this welding process is much efficient as compared to the other welding techniques. Electron beams with the energy between 30 keV and 200 keV are employed for electron beam welding applications. The depth of the weld that is required decides the power of the beam. Weld depths of the order of 300 mm can be achieved using electron beams of sufficiently high power. In the case of large objects, welding can be performed by bringing the electron beam out of vacuum into the atmosphere. With such equipment very large objects can be welded without huge working vacuum chambers.

3.2.3 *Photon–surface interactions*

3.2.3.1 Photoelectric Emission

The photoelectrons are released upon absorbing electromagnetic radiation, such as light. The maximum kinetic energy K_{emax} of an ejected electron is given by

$$K_{emax} = h\nu - W \tag{3.35}$$

where h is the Planck constant and is ν the frequency of the incident photon. W is the work function, which corresponds to the minimum energy required to remove a delocalized electron from the surface of the metal and is given by

$$W = h\nu_0 \tag{3.36}$$

where ν_0 is the threshold frequency for the metal. The maximum kinetic energy of an ejected electron is given by

$$K_{emax} = h(\nu - \nu_0) \tag{3.37}$$

3.2.3.2 X-Ray Photoelectron Emission

When X-rays are directed to the surface, the energy of the X-ray photon is absorbed completely by a core electron of an atom. The core electron then escapes from the atom and is released from the surface if the X-ray photon energy is sufficiently high. The kinetic energy equations (3.36) and (3.37) apply also for this emission. Figure 3.11 illustrates X-ray induced photoelectron emission from the surface.

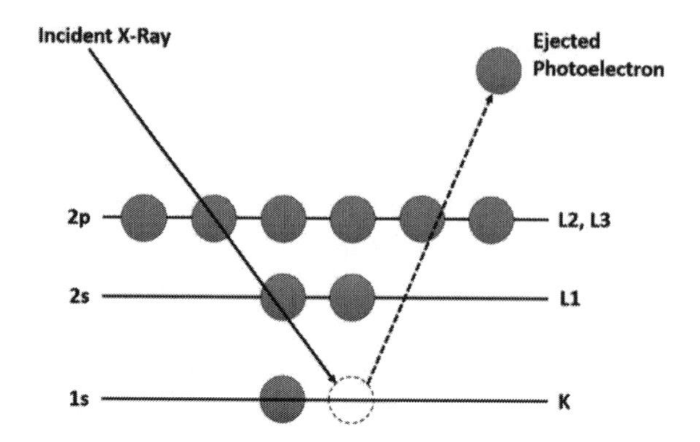

Fig. 3.11: Photoelectron emission from the solid surface caused by incident X-ray.

3.2.4 Electric Field–Surface Interaction

3.2.4.1 Field Emission

Application of an intense electric field (~10^7 V/cm) at a metal surface distorts the potential barrier at the surface in such a way that the electrons from the Fermi sea striking this barrier will have a finite probability of tunneling through for emission rather than escaping over it as in thermionic or photoemission. The effect is quantum-mechanical. The wave function of an electron does not vanish at the classical turning point, but decays exponentially into the barrier where the electron's total energy is less than the potential energy, thereby leaving a finite probability that the electron is found outside of the barrier. For a metal at low temperature, the process can be understood in terms of the illustration shown in Fig. 3.12.

The metal is considered as a potential box, filled with electrons to the Fermi level, lying below the vacuum level by several electron volts. The difference in the energy levels between Fermi and vacuum

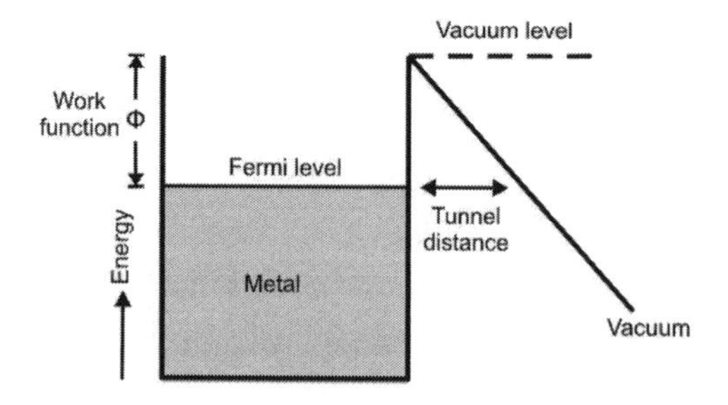

Fig. 3.12: Field emission from a metal at absolute zero temperature (*CC BY-SA 3.0 – https://en.wikipedia.org/wiki/File:Fig216_Field_Emission.PNG*).

levels is the work function Φ. The vacuum level corresponds to the potential energy of an electron at rest outside the metal, in the absence of an external field. In the presence of a strong field, the potential outside the metal is deformed along the line *AB*, so that a triangular barrier is formed. The field electrons can tunnel through this barrier with the most of the electrons escaping from the vicinity of the Fermi level where the barrier is thinnest.

Nordheim[84] has established that the current density *j* in A/cm^2 is given by

$$J = \frac{1.57 \times 10^{-6} \varepsilon^2}{\varphi t^2(x)} \times \exp\left[\frac{-6.83 \times 10^7 \varphi^{\frac{3}{2}} f(X)}{\varepsilon}\right] \tag{3.38}$$

where ε is the electric field in V/cm; $e\varphi$ is the work function in eV.

The functions $t(x)$ and $f(x)$ with the argument $x = 3.79 \times 10^{-4} \sqrt{\varepsilon}/\varphi$ from the value provided by Good and Muller[85].

3.2.5 Thermionic Emission

In thermionic emission, the charge carriers escape from a surface over the potential-energy barrier as a result of thermal excitation due to high temperatures. The carriers (electrons or ions) overcome the work function of the metal. A free electron attempting to leave a conductor experiences a strong force F_x attracting it backwards due to an image charge given by

$$F_x = \frac{-e^2}{4\pi\varepsilon_0(2X)} \tag{3.39}$$

where X is the distance of the electron from the interface; e is the charge on the electron.

With increase in the temperature of the metal, the electrons will move faster and some will have enough energy to overcome the image-charge force and escape. The higher the temperature, the larger is the current of escaping electrons. This temperature-induced electron flow is called thermionic emission. The precise relationship between the voltage and the resulting current flow is called Child's law or the Child–Langmuir law.

Richardson derived an equation for the thermionic emission of electrons as a function of temperature of the emitter assuming that the electrons in a metal behave similar to a perfect gas and follow a Maxwellian distribution of velocities. Richardson's equation is

$$j = aT^2 e^{-b_0/T} \tag{3.40}$$

where j is the saturation current density, T is the temperature of the emitter, and a and b are the constants of the material

Dushman[86] further modified Richardson's equation as

$$j = AT^{\frac{1}{2}} e^{-b_0/T} \tag{3.41}$$

where b_0 is a constant of the emitting surface such that

$$b_0 k = \phi_0 e \tag{3.42}$$

where ϕ_0 is the thermionic work function, e is the electron charge and k is Boltzmann's constant.

3.2.6 Emission From Contact Surfaces in Vacuum Arcs

Cathode spots[87] and anode spots[88] occur on the contact surfaces in certain conditions including arc currents of several thousand amperes, in the presence of electrical arcs in vacuum. Cathode spots are regions of intense ionization, high current density, high power density and high mobility on the cathode surface. In addition to the metal vapour that is continually evaporated from cathode spots, a significant amount of emission of electrons, positive metal ions, excited metal atoms and radiation takes place from the cathode spots. Certain arcing conditions give rise to a single and grossly evaporating anode spot. The temperature of the anode spot is near the boiling point of the anode material. The anode spot serves as a copious supply of metal vapour and energetic metal ions.

3.2.7 Evaporation–Sublimation–Vapour Pressure

The vapour pressure of a liquid or solid is the equilibrium pressure of a vapour above its surface. It corresponds to the pressure of the vapour resulting from evaporation of a liquid or solid in a closed container. As a solid or a liquid evaporates in a closed container, the molecules cannot escape. Some of the gas molecules will eventually return back to the condensed phase. When the rate of condensation of the gas equals the rate of evaporation of the liquid or solid, the gas in the container is in thermodynamic equilibrium with the condensed phase.

The pressure exerted by the gas in equilibrium with a solid or liquid in a closed container at a given temperature is called the vapour pressure. Vapour pressure increases with the temperature of the solid or liquid.

The relation between pressure P, enthalpy of vapourization, ΔH_{vap}, and temperature T is given by,

$$P = A\exp\left(\frac{-\Delta H_{vap}}{RT}\right) \quad (3.43)$$

where $R = 8.3145\ \text{J} \cdot \text{mole}^{-1} \cdot \text{K}^{-1}$ and A are the gas constant and the unknown constant. This is known as the Clausius–Clapeyron equation. If P_1 and P_2 are the pressures at two temperatures T_1 and T_2, the equation has the form:

$$\ln\left(\frac{P_1}{P_2}\right) = \left[\frac{\Delta H_{vap}}{R}\right]\left[\left(\frac{1}{T_2}\right) - \left(\frac{1}{T_1}\right)\right] \quad (3.44)$$

Using the Clausius–Clapeyron equation, it is possible to estimate the vapour pressure at another temperature, if the vapour pressure is known at some temperature, and if the enthalpy of vaporization is known.

The variation of vapour pressure[89] of selected metal elements with temperature is shown in Fig. 3.13.

3.2.8 *Vacuum Evaporation*

Vacuum evaporation is the process in which the liquid is made to evaporate at a lower temperature than normal when the pressure above the liquid surface is reduced below the vapour pressure of the liquid. This process is employed particularly for the storage of food products for longer periods without degeneration because the water can be effectively removed at much lower temperatures as the boiling point of water is reduced at low pressure. In the vacuum distillation process, this method is used as the more volatile liquid can be removed fast by vacuum evaporation.

The vacuum freeze drying process is used to preserve perishable items. In this process, the items such as blood plasma are frozen and

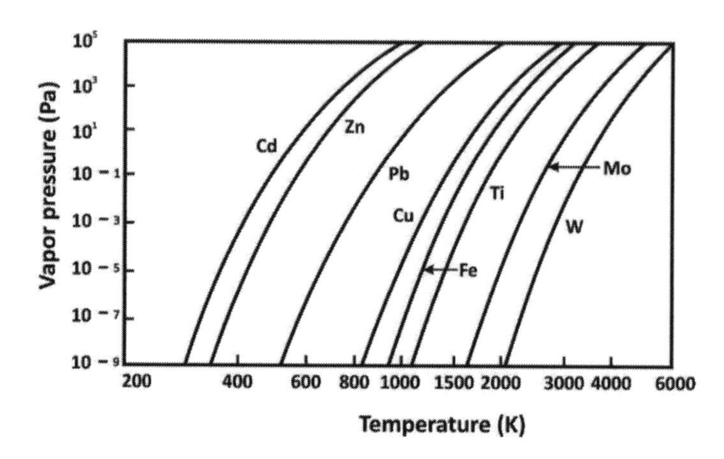

Fig. 3.13. Vapour pressure of selected metals.

the water content in the items is removed by sublimation directly from the solid phase to the gas phase by controlled heating under the vacuum condition.

Physical vapour deposition can be achieved by vacuum evaporation. In this process, thin[90] or thick[91] films of material are deposited onto solid surfaces. The technique consists of heating a material in a vacuum chamber to produce a flux of vapour that deposits the material onto a surface. The material to be vapourized is typically heated until its vapour pressure is high enough to produce a rate of deposition that is higher than the impingement rate of the residual gas by a few orders of magnitude. This can be achieved by using an electrically resistive heater or by high-energy (a few keV) electron bombardment of the substance to be deposited.

References

1. P. A. Redhead, J. P. Hobson and E. V. Kornelsen, Adv. Electron. Electron Phys. **17**, 323 (1962).
2. V. O. Altemose, J. Appl. Phys. **32**, 1309 (1961).
3. B. B. Dayton, 1959 Vac. Symp. Trans. 101 (1960).
4. H. L. Eschbach, F. Gross and S. Schulien, Vacuum **13**, 543 (1963).
5. H. A. Steinherz, "Handbook of High Vacuum Engineering", Reinhold Publishing Corporation, New York (1963).
6. V. O. Altemose, Symp. Art of Glass-Blowing 7, Am. Sci. Glass-Blower Soc. Wilmington, Del. (1962).
7. D. Alpert and R. Buritz, J. Appl. Phys. **25**, 202 (1954).
8. E. Serra, A. Calza Bini, G. Cosoli and L. Pilloni. J. Am. Ceramic Soc. **88**, Issue 1, 15 (2005).
9. R. M. Roberts, T. S. Elleman, H. Palmour and K. Verghese, J. Am. Ceramic Society

62, 495 (1979).
10. F. J. Norton, Trans. 8th Natl. Vac. Symp. USA, 8 (1962).
11. J. E. Lenard-Jones, Trans. Faraday Soc. **28**, 333 (1932).
12. J. Bardeen, Phys. Rev. **58**, 727 (1940).
13. H. Margenau and W. G. Pollard, Phys. Rev. **60**, 128 (1941).
14. B. M. W. Trapnell, "Chemisorption", Academic Press, Butterworths Scientific Publications, New York, London (1955).
15. J. H. De Boer, "The Dynamical Character of Adsorption", Oxford:Clarendon Press (1953).
16. R. Gat and J. C. Angus, J. Appl. Phys. **74**, 5981 (1993).
17. G. Ehrlich, Ann. N. Y. Acad. Sci. **101**, 722 (1963).
18. D. Lee, H. Tomaschke and D. Alpert, Trans. 1961 Natl. Vac. Symp. 8, 153, (1962).
19. D. Alpert, Le Vide, No. 17, **19**, January (1962).
20. J. Frenkel, Z. Physik **26**, 117 (1924).
21. H. Yildirim Erbil, "Surface Chemistry of Solid and Liquid Interfaces", Blackwell Publishing, (2006).
22. I. Langmuir and K. H. Kingdon, J. Chem. Phys. **1**, 3 (1933).
23. S. Brunauer, R H. Emmett, and E. Teller, 7. Am. Chem. Soc. **60** (1938).
24. H. Freundlich, "Colloid and Capillary Chemistry", Methuen and Co. Ltd., London, 111 (1926).
25. K. S. Walton and R. Q. Snurr, J. Am. Chem. Soc. **129** (27), 8552 (2007).
26. H. Steininger, S. Lehwald and H. Ibach, Surface Science **123**, 264 (1982).
27. D. A. King and M. G. Wells, Surface Science **29**, Issue 2, 454 February (1972).
28. P. Hobson, Trans. Natl. Vac. Symp. USA 1961, 26 (1962).
29. P. A. Redhead, Vacuum **12**, Issue 4, 203 (1962).
30. J .P. Cowin, D.J. Auerbach, C. Becker, and L. Wharton, Surface Science **78**, Issue 3, 545 (1978).
31. S. M. George, A. M. De Santolo and R. B. Hall, Surface Science **159**, Issue 1, L425 (1985).
32. M. L. Knotek and P. J. Feibehnan, Phys. Rev. Letters **40**, 964 (1978).
33. D. Menzel and R. Gomer, J. Chem. Phys. **41**, 3311 (1964).
34. P. A. Redhead, Can. J. Phys. **42**, 886 (1964).
35. M. Bender, H. Kollmus, W. Assmann, Proc. EPAC 2006, Edinburgh, Scotland.
36. E. Mahner, "Review of heavy-ion induced molecular desorption studies for particle accelerators". CERN, Technology Department, Vacuum, Surfaces and Coatings Group 27 May, 2011.
37. K. Fukutani, M. B. Song and Y. Murata, Faraday Discuss. **96**, 105 (1993).
38. J. A. Venables, G. D. T. Spiller and M. Hanbucken, Reports on Progress in Physics **47** Number 4, 399 (1984).
39. B. von Lewis and J. C. Anderson, " Nucleation and Growth of Thin Films", Academic Press, New York, 1978.
40. R. V. Stuart and G. K. Wehner, J. Appl. Phys. **35** (6), 1819 (1964).
41. O. Auciello, S. Chevacharoenkul, M. S. Ameen and J. Duarte, J. Vac. Sci. Technol. **A9**, 625 (1991).
42. Ed. K. Kazmerski, "Polycrystalline and Amorphous Thin Films And Devices", Academic Press (1980).
43. S. Alfihed, M. Hossain, A. Alharbi, A. Alyamani and F. H. Alharbi, Journal of Materials **4**, (2013).
44. B. A. Joyce, Rep. Prog. Phys. **48**, 1637 1985).
45. A. Y. Cho, J. Appl. Phys. **41**, 2780 (1970).

46. I. Langmuir and K. H. Kingdon, Phys. Rev. **34**, 129 (1920).
47. M. Kaminsky, “Atomic and Ionic Impact Phenomena on Metal Surfaces”, Academic Press Inc., New York (1965).
48. Z. Sroubek, International Journal of Mass Spectrometry and Ion Physics **53**, 20, 289 (1983).
49. J. F. Wendt and A. B. Cambel, J. Appl. Phys. **34**, 176 (1963).
50. G. M. McCracken, Rep. Prog. Phys. **38**, 241 (1975).
51. G. Carter, J. Vac. Sci. Technol. **7**, 31 (1970).
52. H. D. Hagstrum, Phys. Rev. **104**, 672 (1956).
53. H. S. Massey and E. H. S. Burhop, “Electronic and Ionic Impact Phenomena”, Oxford: Clarendon Press (1952).
54. D. Chattarji, “The Theory of Auger Transitions", Academic Press, London, (1976)
55. H. Bruining, “Physics and Applications of Secondary Electron Emission”. McGraw Hill Book Co. Inc., New York (1954).
56. K. J. Close and J. Yarwood Br. J. Appl. Phys. **17**, 1165 (1966).
57. P. K. Naik, V. G. Kagal, S. L. Verma and S. P. Mhaskar, Appl. Phys. **23** (4) 373 (1980).
58. E. V. Kornelsen, Can. J. of Phys. **42** (2), 364, (1964).
59. E. V. Kornelsen, J. Vac. Sc. Technol. **9**, 624 (1972).
60. E. V. Kornelsen, J. Vac. Sc. Technol. **6**, 173 (1969).
61. E. V. Kornelsen and M. K. Sinha, J.Appl. Phys. **39** (10), 4546 (1968).
62. S. P. Mhaskar and P. K. Naik, Phys. Stat. Sol. (b) **107**, 99 (1981).
63. S. P. Mhaskar and P. K. Naik, Phys. Stat. Sol. (b) **105**, 685 (1981).
64. S. P. Mhaskar and P. K. Naik, Phys. Stat. Sol. a **72**, 15 (1982).
65. S. M. Rossnagel, “Handbook of Vacuum Science and Technology” Ed. D. H. Hoffman, B. Singh and J. H. Thomas III, Academic Press, (1998).
66. G. K. Wehner, Advances in Electronics and Electron Physics **7**, 239 (1955).
67. P. Sigmund, Pan Stanford Publishing Review Volume, April **13**, (2011).
68. D. A. Thompson, Radiat. Effects **56**, 105 (1981).
69. K. L. Merkle and W. Jäger, Phil. Mag. A **44** 741 (1981).
70. H. Yamaguchi, A. Shimase, S. Haraichi and T. Miyauchi, J. Vac. Sci. Technol. B **3**, 71 (1985).
71. J. Lindhard, M. Scharff and H. E. Schiott, Mat. Fys. Medd. Dan. Vid. Selsk. **33**, No. 14 (1963).
72. C. C. Rousseau, W. K. Chu and D. Powers, Phys. Rev. A**4**, 1066 (1970).
73. W. D. Wilson, L. G. Haggmark and J. P. Biersack, Phys. Rev. **15B**, 2458 (1977).
74. S. Aisenburg, J. Vac. Sci. Technol. **10** , 104 (1973).
75. D. M. Mattox, and V. H. Mattox (editors) "50 Years of Vacuum Coating Technology and the Growth of the Society of Vacuum Coaters" Society of Vacuum Coaters (2007).
76. A. J. Dekker, "Secondary Electron Emission", vol. 6, ed. F. Seitz, D. Turnbull, H. Ehrenreich. Academic Press, NY (1958).
77. W. S. M. Werner, Surf. Interface Anal. **31**, 141, (2001).
78. H. Bruining, “Physics and Applications of Secondary Electron Emission”, McGraw Hill Book Co., Inc. New York (1954).
79. R. E. Simon, Nuclear Science, IEEE Trans. on Nuclear Science **15**, (3), 167 (1968).
80. P. G. Slade, “ The Vacuum Interrupter, Theory, Design and Application”, CRC Press, Taylor & Francis Group, Boca Raton-London-New York (2007).
81. R. F. Bunshah and M. A. Cocca, “Electron Beam Melting, Annealing and Distillation Techniques of Metal Research”, 1 Part 2, 717 Interscience Publishers, New

York (1968).
82. D. Reuter, A. D. Wieck and A. Fischer,, Rev. Sci. Instrum. **70**, 3435 (1999).
83. H. Schultz, "Electron Beam Welding", Abington Publishing, Woodhead Publishing Ltd., Cambridge, England (1993).
84. L. W. Nordheim, Proc. Roy. Soc. (London) A **121**, 626 (1928).
85. R. H. Good and E. W. Muller, "Field Emission at Higher Temperatures", Handbuch der Physik **21**, 190 (1956).
86. S. Dushman, Phys. Rev. **21**, 623 (1923).
87. B. Juttner, J. Phys. D: Appl. Phys. **24**, R103 (2001).
88. H. C. Miller, IEEE Trans. on Plasma Science **11** (2) 76, (1983).
89. R. E. Honig and D. A. Kramer, RCA Rev. **30**, 285 (1969).
90. K. L. Chopra, "Thin Film Phenomena", Krieger Pub. Co. (1979).
91. R. F. Bunshah, Metallurgical Coatings, Volume I: Proc. International Conf. (1978).

4

Interaction of Neutrals, Charged Particles and Radiation with Gases in Vacuum

The subject of interaction of neutrals, charged particles, radiation, electric and magnetic fields with gases in vacuum is discussed as this subject has a significant impact on various physical and chemical processes in vacuum.

4.1 Disassociation, Excitation and Ionization

Collisions with gas particles can be of the elastic type wherein momentum and energy are preserved. Inelastic interactions that involve the transfer of energy can result in

a. Dissociation involving breaking up of a molecule into atoms provided the energy exceeds the molecular binding energy.
b. Excitation involving pushing an electron in an atom/molecule into a higher energy excited state. The excited state can be of extremely short duration (less than a nanosecond) as the excited electron jumps back to the original energy state while releasing radiation.
c. Ionization involving knocking off an electron from the atom/molecule provided the energy of the striking incident electron exceeds the binding energy of the electron in the atom/molecule. Ionization can also result from electrons having the energy less than the minimum energy required for ionization if an excited atom/molecule is struck by such electrons. The formation of negative ions occurs as a result of the attachment of electrons to

the neutral gas particles. High energy neutrals, positive ions, beta rays and radiation also can cause ionization. Thermal ionization can occur at high temperatures as in the case of high-temperature flames.

4.1.1 Mean Free Path of Electrons and Ions

The average distance λ_e traveled by an electron between two successive collisions with gas particles is the mean free path of electrons[1] and is given by

$$\lambda_e = 4\sqrt{2}\lambda_g \tag{4.1}$$

where λ_g is the mean free path (MFP) of the gas in which the electron is moving.

The maximum energy that the electron gains between two successive ionizing collisions in electric field E is given by

$$eE\lambda_e = \frac{eV\lambda_e}{d} \tag{4.2}$$

where d is the distance between the electrodes and e is the electron charge. Thus, for the electron-impact ionization of an atom, it is necessary that the energy $eE\lambda_e$ of the electron exceeds the binding energy of the electron in the atom.

The MFP λ_i of a positive ion in the gas in which it is moving is given by

$$\lambda_i = \sqrt{2}\lambda_g \tag{4.3}$$

4.1.2 Ionization Potential and Ionization Efficiency/Cross Section

The ionization potential corresponds to the minimum energy required to dislodge the weakest bound electron from its normal state into external space. Values of ionization potentials U_i of typical gases and metal atoms taken from various sources[1,2,3] are given in Table 4.1.

The ionization efficiency of electrons corresponds to the number of ion pairs produced per unit length and per unit pressure. Figure 4.1 shows variation of ionization efficiency with electron energy. Here, the electron energy is given in terms of eV (electron volts). The energy of 1 eV is gained by an electron across a potential difference of 1 Volt in an electric field. The ionization efficiency

Table 4.1. Ionization potential of certain gases and metal ions

Gas	Ionization Potential (U_i)
Air	14.0
A	15.76
CO_2	14.4
H	13.6
N	14.53
O	13.62
C	11.26
Cu	7.72
Ag	7.57
Cr	6.76
W	7.86
Bi	7.28

curve exhibits regions of maxima for the gases for electron energies spreading from a few eV to few hundreds of eV. This variation presents an important observation for engineers and scientists engaged in vacuum practice. It is necessary to consider variation of the ionization efficiency for understanding the design aspects of ionization gauges for the measurement of vacuum (pressure) and of ion pumps for the production of high vacuum.

4.1.3 Positive Ion-Impact Ionization

High-energy positive ions can cause ionization of gas atoms on collision. Engel and Steenbeck[4] have studied ionization coefficients for alpha particles in air. Rudd et al[5] have measured cross-sections for the production of positive and negative charges for proton impact at 5–4000 keV on He, Ne, Ar, Kr, H_2, N_2, CO, O_2,CH_4 and CO_2.

4.1.4 Ionization by Photons and Thermal Ionization

Ionization by photons can be effected through the photoelectric effect and the Compton effect. Such interactions will cause the ejection of an electron from an atom at relativistic speeds, in the form of a beta particle that can cause ionization of other atoms. In general, thermal

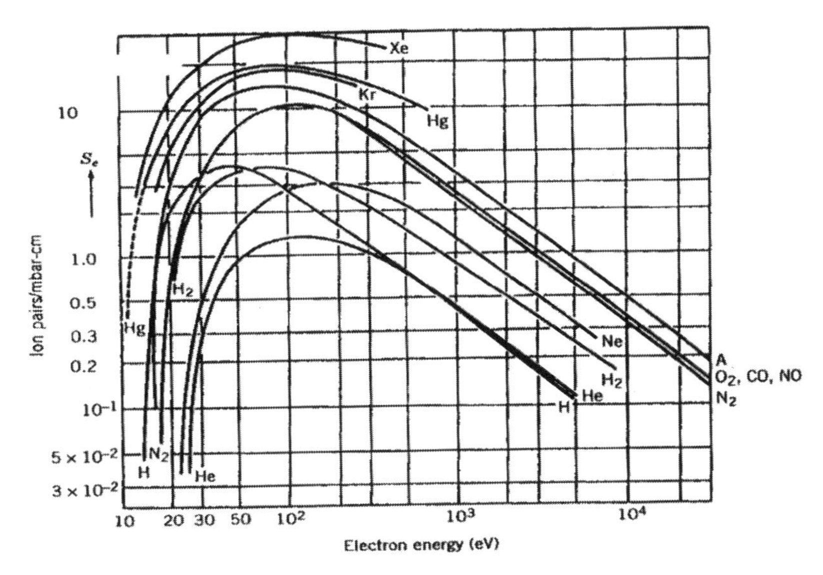

Fig. 4.1. Variation of ionization efficiency with electron energy (*Reproduced from A. Von. Engel, 'Ionized Gases', AVS Classics Series, with the permission of the American Vacuum Society*).

ionization refers to the ionization caused by molecular collisions, electron collisions and radiation at high temperatures. This type of ionization occurs in high-temperature flames. High velocities associated with gas atoms at high temperature can result in the atoms gaining sufficient energy to cause ionization upon impacts as the electrons, normally bound to the atom in orbits around the atomic nucleus will be ejected from the atom. An electron gas is thus formed that co-exists with the gas of atomic ions and neutral atoms. This state of matter is called a plasma. The Saha equation[6,7] describes the degree of ionization of this plasma as a function of the temperature, density, and ionization energies of the atoms.

4.1.5 Deionization

The ionized gas rapidly returns to the electrically neutral state upon withdrawal of the agencies causing the ionization, as a result of recombination of oppositely charged ions and due to the diffusion to the walls.

4.2 Electrical Breakdown in Vacuum

Knowledge of the phenomenon of electrical breakdown in vacuum is important as it explains the performance and limitations of many vacuum devices. It is important to understand that there are two different types of electrical breakdown processes in vacuum. These are commonly known as

(a) Gas breakdown
(b) Vacuum breakdown

4.2.1 Gas Breakdown

The gas breakdown is not necessarily limited to vacuum environment. It can also occur at atmospheric and higher pressures.

The Townsend discharge theory[8] forms the basis of the gas breakdown. Pairs of charged particles, comprising positive ions of the residual gas and the electrons, are present in the atmospheric air due to the stray ionizing agencies. The theory states that a small number of free stray electrons and positive ions, accelerated by an electric field, give rise to electrical conduction through a gas by an avalanche process. A Townsend avalanche is a cascade reaction involving electrons in a region within an electric field. The positive ions generated by electron–impact ionization drift towards the cathode, while the free electrons drift towards the anode. Consider one such electron that will undergo an ionizing collision with another atom/molecule of the medium. The two free electrons then travel together some distance before another collision occurs. The number of electrons travelling towards the anode is multiplied by a factor of two for each collision, so that after n collisions there are 2^n free electrons. Progressively, this results in an electron avalanche.

Electrons are also emitted by other processes such as photoelectric emission and secondary electron emission due to positive ion bombardment of the cathode.

α, called the first Townsend ionization coefficient, corresponds to the number of ionizing collisions per unit length in the direction of the field. It is given by

$$\alpha = Ap\exp\left(\frac{-AU_i}{E/p}\right) \tag{4.4}$$

where A is a constant, U_i is the ionization potential of the gas, p is the gas pressure and E is the electric field.

If n_0 number of electrons per unit area are released per unit time by the cathode in space between two electrodes located at a distance

d apart, then the number n_1 of electrons arriving at the anode per unit area per unit time is given by

$$n_1 = n_0 e^{\alpha d} \tag{4.5}$$

on converting to current, this is

$$I_1 = I_0 e^{\alpha d} \tag{4.6}$$

where I_1 and I_0 are currents corresponding to n_1 and n_0 respectively.

As the current increases due to the ionization process, secondary electrons are then released from the positive ion bombardment of the cathode and by photoelectric process.

γ, called the second Townsend ionization coefficient, is the number of new electrons released from the cathode for each positive ion incident. It is established that U_b, the breakdown voltage of the gas with an ionizing potential U_i, is given by the equation

$$U_b = \frac{-AU_i(pd)}{\ln\left[\dfrac{A(pd)}{\ln(1/\lambda)}\right]} \tag{4.7}$$

where A is a constant. Thus, the breakdown voltage of gas with a given ionizing potential is a function of the product of the gas pressure and the inter-electrode distance.

This is also expressed in the form

$$U_b = \frac{a(pd)}{\ln(pd) + b} \tag{4.8}$$

and

$$U_b = fn(nkTd) \tag{4.9}$$

where n is the gas density. This is the statement of Paschen's law[9]. It can be seen that the Paschen's curve in Fig. 4.2 has a minimum value of breakdown voltage U_b at a particular product pd value for a gas. For air, the minimum value of U_b is about 327 V corresponding to the pd value of about 5.7 Torr · mm (7.59×10^{-1} Pa · m).

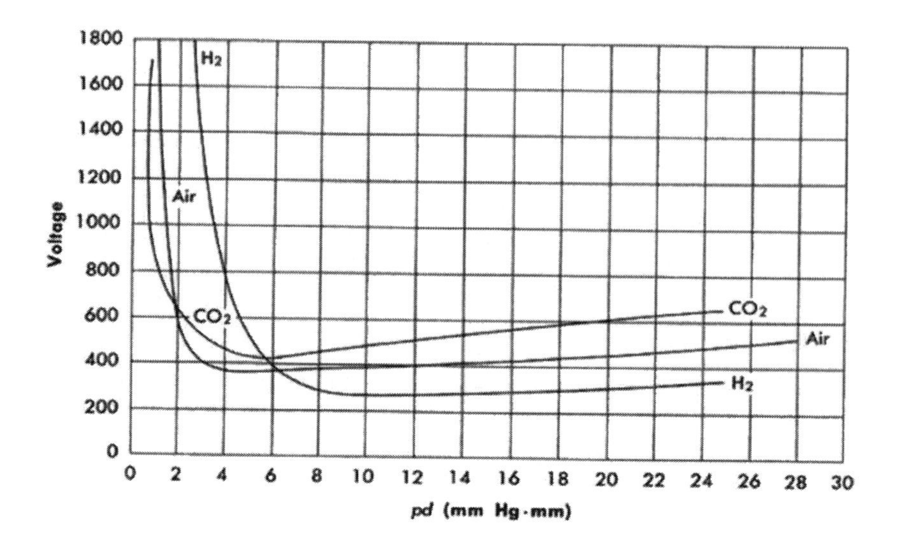

Fig. 4.2. Paschen's curve (*Warfare Centres, NSWC Crane Division, Naval Sea Systems Command (NAVSEA)- U.S. Navy http://www.navsea.navy.mil/Home/Warfare-Centers/NSWC-Crane/Resources/SD-18/ Products/Connectors/Derating/*).

Voltage–current characteristics of the vacuum gap

If a current limiting resistance is used in the circuit that includes the vacuum gap, the current remains self-sustained by the avalanche after the breakdown, but with a lower voltage. Figure 4.3 illustrates different regions of the voltage–current characteristics[1,2] as explained below.

Region AB: Shows the gradual increase in voltage with current as the electron-positive ion pairs formed resulting from stray ionization caused by cosmic radiation, are collected at the anode and the cathode respectively.

Point B: Current saturation occurs at this point

Region BCD: With further increase in voltage, ionization occurs as a result of impact with gas particles with electrons that have acquired sufficient energy. As the current increases due to the ionization process, secondary electrons are released from the positive ion bombardment of the cathode. These additional electrons along with those released by photoelectric process cause the current to increase much more rapidly to finally cause breakdown.

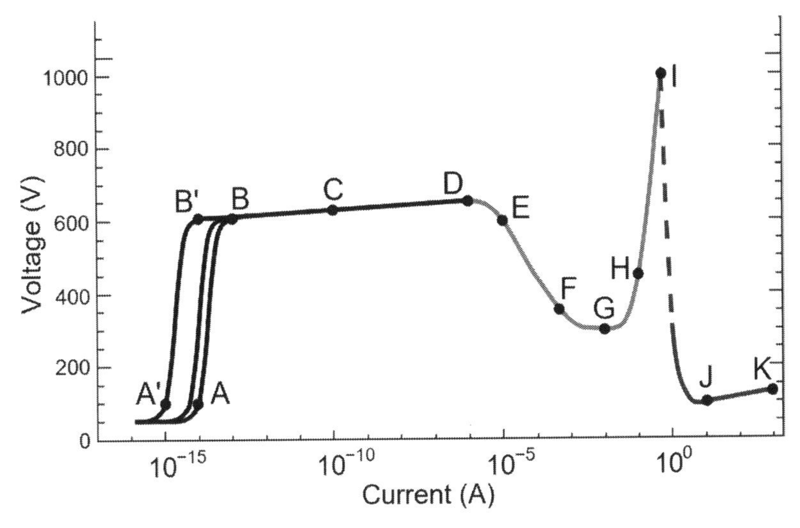

Fig. 4.3. Voltage–current characteristics of Townsend discharge (*CC BY-SA 3.0 -Wikigian & Chetvorno (Wikimedia Commons) – https://commons.wikimedia.org/wiki/File:Glow_discharge_current-voltage_curve_English.svg*).

Point D: The Townsend breakdown occurs at this point. The region *AD* is called ‘dark discharge‘ region as it is characterized by very low current and absence of luminous glow.

Region EF: The mode of the discharge after the Townsend breakdown is governed by the value of the resistance in the circuit. If the resistance is low, the discharge can results in an arc. If the resistance is sufficiently high, a low current glow luminous discharge is established as the voltage drops to point *E* and further to point *F*.

Region FG: This region corresponds to ‘normal glow’ discharge in which the current density is constant. This means that the current increases proportionately with the cross sectional area of the glow.

Region GH: In this region, the joule heating of the cathode further enhances the electron emission giving rise to ‘abnormal glow’ discharge and increase in current density.

Region HI: This region corresponds to unstable region of glow-to-arc transition as the current keeps on increasing till an avalanche breakdown occurs at point *I*.

Region JK: This corresponds to the arcing region, characterized by high current, low voltage and a highly bright arc discharge.

4.2.2 Glow Discharge

Most of the initial studies of gas discharges in vacuum have been conducted in 'discharge tubes'. A typical discharge tube has cylindrical blocks of metal electrodes sealed at the two ends of the tube which is connected to a vacuum pump. The electrodes serve as anode and cathode by maintaining them at appropriate DC potentials using a power supply with a resistance in series to limit the circuit current. If pressure in the tube is maintained close to 10^4 Pa with an applied voltage of about 1 kV across the inter-electrode gap of about 30 cm, a uniform glow appears throughout the gap. As the pressure is lowered, the glow exhibits alternate bright and dark regions. The bright and the dark regions are termed as striations. The dark spaces are named after the scientists in the field of gas discharges. Regions of glow represent much more active ionization and excitation as compared to the regions of dark spaces. The colours of the glow regions vary with the gas and are different at different locations for the particular gas. The variety of colors associated with glow discharges finds application in the field of illumination.

4.2.3 Vacuum Breakdown

Unlike the gas breakdown which is predominantly a process involving electron avalanche in gas, the vacuum breakdown is initiated by surface related phenomena involving field emission of electrons from the surface under the influence of high electric field. The subject of the vacuum breakdown has been reviewed by Farall[10] and Latham[11] separately.

Electric field

The magnitude of electric field E is defined as the force F experienced by a unit electric charge q given by

$$E = \frac{F}{q} \tag{4.10}$$

The electric field is a vector field with SI units of Newtons per Coulomb ($N \cdot C^{-1}$) or Volts per meter ($V \cdot m^{-1}$). The SI base units of the electric field are $kg \cdot m \cdot s^{-3} \cdot A^{-1}$. The direction of the field is given by the direction of that force.

Alpert[12] et al first suggested that the electric field E across a gap d is given by

$$E = \beta \frac{V}{d} \tag{4.11}$$

where V is the voltage across the gap d. β is considered as the field enhancement factor. Further, β is considered as a product of two components β_m, influenced by the microscopic surface conditions such as projections and β_g, influenced by the non-uniform field distribution. Thus,

$$\beta = \beta_m \beta_g \tag{4.12}$$

Field emission

Field emission is primarily responsible for the initiation of vacuum breakdowns as the field electrons are released from one or more micro-projections on the cathode surface when exposed to a high electric field. The field emission of electrons is discussed in section 3.2.4.1. The quantum mechanical field theory for field emission under intense electric fields was first proposed by Fowler and Nordheim[13]. According to the theory, the electrons from the solid surface would tunnel across the potential barrier at the surface and escape into vacuum. These electrons

- can evaporate metal from the anode
- can also evaporate the cathode by joule heating of the emitting region
- can evaporate the particles released by the strong field[14] from the anode

Electrical breakdown can result from the ionization of the metal vapour. A value of critical breakdown field E_c at which the breakdown occurs exists for a given metal.

Thus

$$E_c = \beta \frac{V_b}{d}$$

where V_b is the breakdown voltage for the gap d. It can be seen that for a given gap d, the breakdown voltage V_b is influenced by the value of the field enhancement factor β for a metal having a given value of critical breakdown field E_c. A gap of 100 mm in vacuum breaks down at 800 kV[15].

Such a breakdown results in the formation of the vacuum arc. Alternatively, an arc can also be established by parting of contacts carrying current. This is explained by the rupture of the molten metal bridge that is formed during the parting of the contacts[15].

The two main modes of vacuum arc are

- The diffuse vacuum arc associated with a diffuse inter-electrode plasma and a diffuse collection of current at the anode.
- The columnar vacuum arc that forms into a single high vapor pressure arc column with cylindrical boundaries. At the highest currents, the electrode regions of the arc column exhibit intense activity with ejection of jets of materials from the electrode faces.

4.3 Plasma

Plasma is a region in which macroscopically the number of positive and negative charges are approximately equal, thus rendering the region electrically neutral. The positive column region in a glow discharge tube and electrical arcs are examples of plasma. The Sun, the stars, the interplanetary and interstellar spaces is filled with plasma. Plasma is considered as a distinct state of matter. Studies of plasma physics have made important contributions to the field of nuclear fusion.

The thermonuclear fusion of deuterium H^2 and tritium H^3 nuclei results in formation of helium nuclei He^4 and release of an enormous amount of energy as can be seen from the equation below.

$$ {}_1H^2 + {}_1H^3 \rightarrow {}_2He^4 + {}_0n^1 + 17.6 \text{ MeV} \tag{4.13} $$

In order to achieve the fusion, it is essential to obtain a temperature above 100 million K and the plasma density higher than 1000 particles per cubic meter. Also it is necessary to avoid impurities of nuclei of high atomic numbers in the plasma as they cause cooling of the plasma. For this reason, it is essential to use ultrahigh vacuum conditions.

Heating of the plasma is achieved by the electric current induced by transformer arrangement and by other additional means. Magnetic confinement fusion devices are utilized to contain the plasma. Ultrahigh vacuum conditions are required to minimize the impurities.

References

1. J. D. Cobine, "Gaseous Conductors -Theory and Engineering Applications", Dover Publications, New York, (1958).
2. P. G. Slade, "Electrical Contacts-Principles and Applications", Marcel Dekker, Inc., New York, Ed. P. G. Slade (1999).
3. Handbook of Chemistry and Physics, 3rd Electronic Edition.
4. A. von Engel and M. Steenbeck, "Elektrische Gasentladungen ihre Physik und Technik" **1**, 44 (1932).
5. M. E. Rudd, R. D. DuBois, L. H. Toburen, C. A. Ratcliff and T. V. Goffe, M. Phys. Rev. A **28**, (6), (1983).
6. M. N. Saha, Philosophical Magazine Series 6 **40**, 238 (1920).
7. M. N. Saha, Proc. Royal Soc. A **99,** 697 (1921).
8. J. S. Townsend, "The Theory of Ionization of Gases by Collision", Constable & Company Limited, London (1910).
9. F. Paschen, Annalen der Physik **273**, (5) (1889).
10. G. A. Farall, IEEE Trans. Electrical Insulation, EI-**20**, 815 (1985).
11. R. V. Latham. "High Voltage Vacuum Insulation: Basic Concepts and Technological Practice", Academic Press, London, 184 (1995).
12. D. Alpert, D. Lee, E. Lyman, H. Tomaschke, J. Vac. Sci. Technol. **1**, 33 (1964).
13. R. H. Fowler and L. W. Nordheim Proc. Royal Soc. **A 119**, 173 (1928).
14. P. G. Slade and M. Nahemow, J. Appl. Phys. **49**, 3290 (1971).
15. P. G. Slade, "The Vacuum Interrupter, Theory, Design and Application", CRC Press, Taylor & Francis Group, Boca Raton-London-New York (2007).

5

Measurement of Pressure

Processes occurring in vacuum are mainly influenced by the order of vacuum. Hence, it is essential to measure the gas pressure with fair accuracy. Gauges commonly used for the measurement of pressure in vacuum practice are discussed here. The selection of the type of gauge is determined by the range of pressure or the order of vacuum.

There are three major groups of gauges. These include mechanical gauges, thermal conductivity-based gauges and ionization gauges.

5.1 Mechanical Gauges

Mechanical gauges measure pressure by making use of the mechanical force when exposed to a difference in pressure. The gauges in this class include

- U-tube manometers
- McLeod gauge
- Diaphragm gauge
- Knudsen gauge
- Bourdon gauge

5.1.1 Manometers

A simple manometer as shown in Fig. 5.1, is a U-shaped tube filled with a low vapour pressure fluid.

One arm of the U-tube is evacuated to high vacuum and sealed while the other arm is exposed to the unknown pressure to be determined. In this case, the difference in the levels of the fluid in two arms gives an indication of the unknown pressure provided the pressure in the sealed-off arm is a few orders of magnitude lower

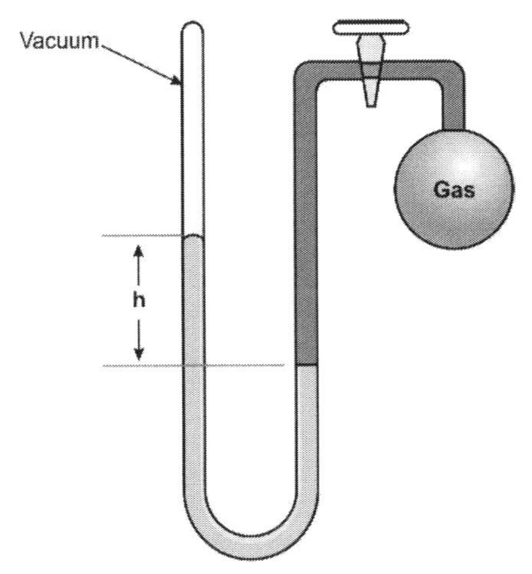

Fig. 5.1. Simple U-tube manometer with one arm sealed at high vacuum.

than the unknown pressure. If h is the difference in levels, then pressure P is given by

$$P = g\rho h \tag{5.1}$$

where g is the gravitational acceleration and ρ is the fluid density. P is expressed in Pa if g and h are taken in MKS units.

Alternatively, one end can be maintained at the atmospheric pressure and the other arm is exposed to the gas the pressure of which is to be measured. In this case the unknown pressure is obtained by subtracting the pressure due to the difference in levels from the atmospheric pressure.

Many modifications have been made in the U-tube manometers by using more accurate methods such as the use of a travelling microscope on a precision travel screw. In another modification[1], a mirror mount is supported on the wall of the enlarged U-tube. The mirror is attached to an arm and floats on the mercury surface. Angular deflection of the mirror is measured as the mercury level changes. Sensitivities of the order of 1×10^{-2} Pa to 1×10^{-1} Pa have been reported using these modifications. The fluids used can be mercury or low vapour pressure fluids such as those used in diffusion pumps. The latter is useful for increasing the sensitivity resulting from increased height of the fluid column due to the density of the fluid which is lower than that of mercury by an order of magnitude. However, the lower density fluid can cause errors due to the solubility of the gas and sluggishness due to the high viscosity of the fluid and its tendency to stick to the glass walls.

5.1.2 McLeod Gauge

The McLeod gauge is based on principle of Boyle's law. A certain amount of the gas at the unknown pressure P_x and of a given volume V_x and temperature is trapped and further compressed. The unknown pressure P_x is then determined by measurement of the pressure P_y and the volume V_y of the compressed gas by the relation as given below.

$$P_x V_x = P_y V_y \tag{5.2}$$

Figure 5.2 illustrates the McLeod gauge which uses mercury. Here, the initial volume V_x includes the volume of the bulb and the volume of the closed end capillary above it. Trapping of the gas in the closed capillary and its further compression is achieved by mechanically raising the level of mercury in the open tube until it is in level with the top of the closed capillary as shown in Fig. 5.2 (a). The two capillaries of equal cross sectional areas a are used. Thus, the pressure P_y of the compressed gas is h which corresponds to the difference in heights of the mercury columns in two capillaries and which also equals the height of the compressed gas. Here, the units of pressure are taken in Torr for convenience as mercury is used and since 1 Torr of pressure corresponds to 1 mm height of the mercury column. Also,

$$V_y = ah \tag{5.3}$$

Thus,

$$P_x V_x = h \times ah = ah^2$$

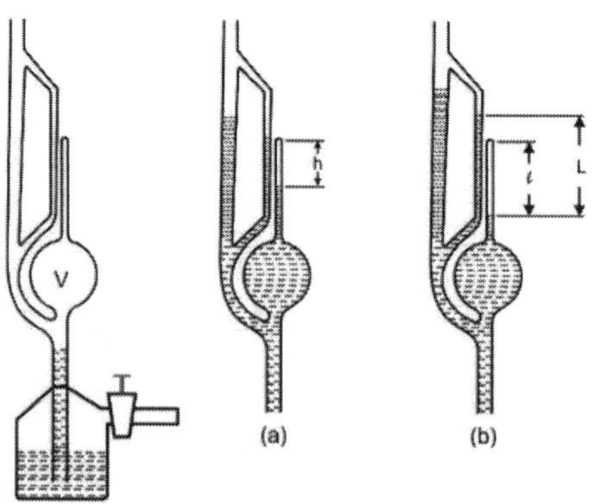

Fig. 5.2. McLeod gauge.

Thus

$$P_x = \frac{ah^2}{V_x} \tag{5.4}$$

where a/V_x is a constant for the gauge. This approach is useful for pressures < 100 Pa as the scale is quadratic.

To cover a wider range of pressure measurements, another method is used in which the mercury is raised to a prescribed level at a distance l from the from the top in the closed capillary as shown in Fig. 5.2 (b) and the difference L in levels of mercury in two capillaries is measured. In this case,

$$P_x V_x = alL, \tag{5.5}$$

It is essential that high-purity mercury is used in the gauge. A liquid-nitrogen trap should be used between the system and the gauge to prevent water vapour and diffusion pump vapors entering the gauge and also to prevent mercury vapours from the gauge entering the system. The walls of the capillaries must be clean.

The McLeod gauge, being an absolute gauge, is used for calibrating other vacuum gauges. The accuracy levels of this gauge change with the pressure as it is less accurate at about 10^{-3} Pa which happens to be the lower limit of measurement, the upper limit being 1.3×10^3 Pa. The gauge is fragile and cannot be used for measurement of changing pressure in the system. The measurements of pressure made with the McLeod gauge are independent of the type of gas, except condensable vapours. It covers a wide range of pressures with a fairly good accuracy. A continuous monitoring of the pressure is not possible with this gauge.

5.1.3 Diaphragm gauges

In one of the versions by Alpert et al[2], a thin metal diaphragm is mounted inside a tube as shown in Fig. 5.3. The difference in pressure across the diaphragm will cause the central region of the diaphragm to deflect. The diaphragm serves as one plate of the capacitor while the other fixed plate is mounted in front of the diaphragm. One side of the gauge is exposed to high vacuum or is sealed off. The capacitance of the two plates is measured and gives indication of the differential pressure. This gauge is also known as capacitance manometer. The capacitance gauge is capable of detecting extremely small diaphragm movements. Commercially available

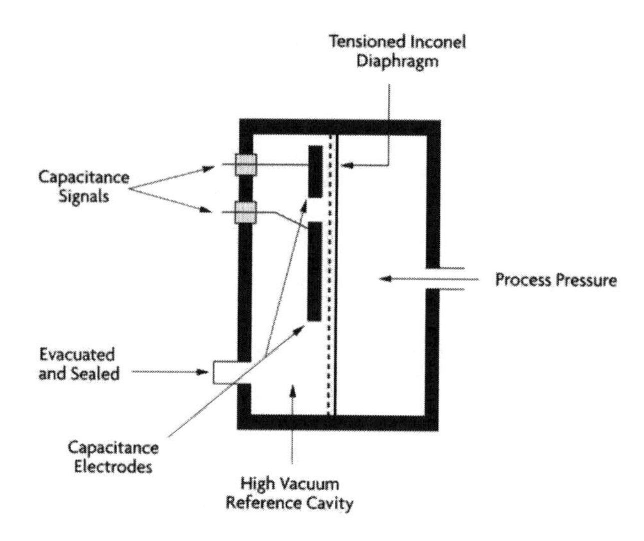

Fig. 5.3. Capacitance manometer (*Copyright 2000 Putnam Media Inc. and Omega Engineering, Inc. Reproduced with permission of Omega Engineering, Inc.*).

diaphragm gauges use Inconel for compatibility with any gas. They offer a useful range of pressure measurement from ~10^{-3}–10^{5} Pa with accuracies (as percentage of reading) of ±0.25% to ±0.5%. The pressure measurement with this gauge is independent of the gas type. Thicker diaphragms can measure pressures in the rough vacuum to the atmospheric range. Two or more capacitance sensing heads can be mounted to cover pressures in a wider range. The capacitance diaphragm gauge is commonly employed in the semiconductor industry, as its Inconel body and diaphragm are corrosion resistant.

5.1.4 Knudsen Gauge

Figure 5.4 illustrates the Knudsen gauge. A rectangular frame with a movable vane V is suspended by a quartz fiber. Further, a mirror is mounted on the fiber to measure changes in pressure. Two fixed heated plates S_1 and S_2 are located in front of the vanes as shown. At low pressure, the momentum of gas molecules rebound from the heated plates is higher than that of the molecules rebound from the cooler vane. This results in the application of a net force on the vane, thereby causing a rotation of the assembly that can be measured by the lamp and scale arrangement.

The expression for the pressure P_i inside the vessel is given by

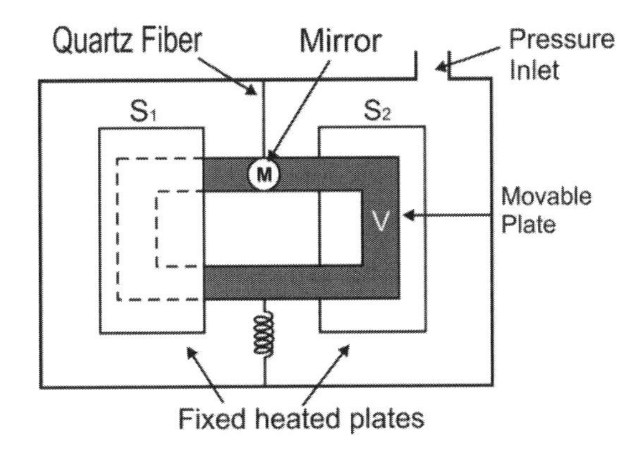

Fig. 5.4. Schematic diagram of Knudsen's gauge.

$$P_i = \frac{KF}{\left[(T_s / T_m) - 1\right]^{\frac{1}{2}}} \tag{5.6}$$

where K is the constant for the gauge, F is the force which can be evaluated from the deflection and the elastic constants of the suspension fiber; T_s is the temperature of the stationary plates; T_m is the temperature of the moving vanes.

The equation is valid provided the mean free path of the gas is larger than the gap between vanes and the fixed plates amounting to the fact that the pressure is less than 1×10^{-1} Pa. The Knudsen gauge offers measurement of pressures in the range 10^{-6} Pa to 10^{-1} Pa.

5.1.5 Bourdon Gauge

The Bourdon gauge is applicable at the high pressure end (6×10^4 Pa upwards) of vacuum. The gauge is shown in Fig. 5.5. The gauge uses an elastic measuring element in the form of a thin-walled tube, bent in a circle with the open end connected to the vacuum system of which the pressure is to be measured. The tube produces a deflection resulting from pressure difference. The deflection is further transformed into an amplified deflection of the pointer on the measuring scale. This gauge also indicates the differential pressure as one end of the gauge is always exposed to atmosphere.

5.2 Thermal Conductivity-based Gauges

Dissipation of heat from a surface can take place by radiation and/

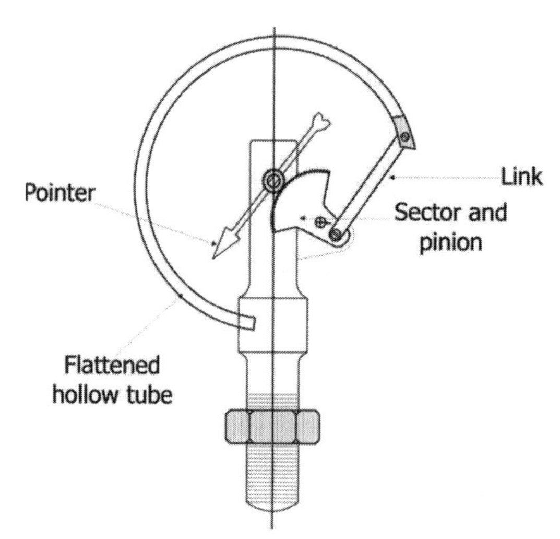

Fig. 5.5. Bourdon gauge.

or by conduction through the gas. When the mean free path of the gas is less than the distance between the hot and cold surfaces, the heat loss is by conduction and depends on the gas density.

In thermal conductivity gauges, a wire element is used that is heated by the passage of electric current. Pirani[3] explained three different modes in which such a gauge can be used. These include observation of

- The current as a function of pressure using constant voltage across the heated element
- The resistance as a function of pressure using constant current passing through the heated element
- The energy required maintain the resistance of the heated element constant as function of pressure.

In the Pirani gauge as shown in Fig. 5.6, a heated metal wire element suspended in gas loses heat as the gas as molecules collide with the wire and remove heat. At lower gas pressure the gas density falls and the heat loss from the wire will be less and the wire will not cool faster. Indication of pressure can be made by measurement of the temperature of the element. The electrical resistance of a wire varies with its temperature. Hence, the resistance and the temperature of the wire is also indicative of the gas pressure. Campbell[4] operated the gauge in the Wheatstone bridge and adjusted the impressed voltage to keep the bridge in balance. He established that the pressure was proportional to the square of the applied voltage. The useful

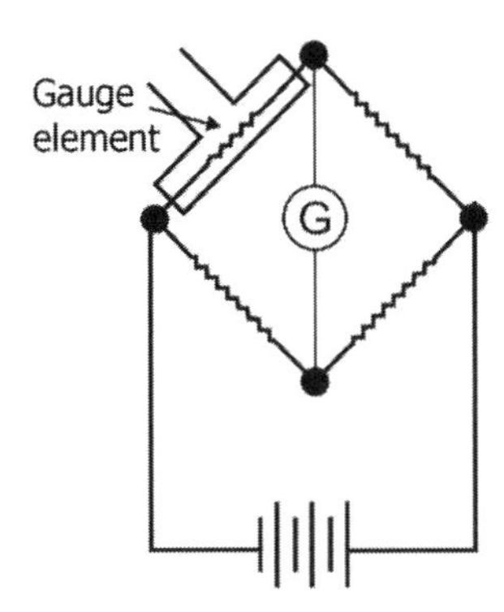

Fig. 5.6. Schematic diagram of Pirani gauge.

pressure measurement range of the Pirani gauge is 1×10^{-1} Pa to 65 Pa. The lower limit of the pressure measurement is governed by the fact that at lower pressure the gas density is too low to cause any significant change in the wire temperature as the heat loss by conduction through gas becomes too small and not measurable. The thermal conductivity and heat capacity of the gas influence the measurement of pressure in this gauge.

In the thermocouple gauge, a constant electrical current is supplied to the filament inside the gauge to which a thermocouple is spot welded for measurement of temperature which is indicative of the gas pressure.

Chang et al[5] reported gas pressure sensor capability of a ZnO nanowire. They studied the growth and characterization of lateral ZnO nanowires on ZnO:Ga/glass templates. The average length and average diameter of the laterally grown ZnO nanowires were 5 μm and 30 nm, respectively. A single nanowire was bridged across two electrodes. By measuring the current–voltage characteristics of the samples at low pressure an increase in currents from 17 nA to 96.06 nA was observed by lowering the pressure from 1.3×10^{-1} Pa to 7×10^{-4} Pa. The authors have claimed that ZnO nanowires prepared in the study are potentially useful for pressure sensing in vacuum.

5.3 Ionization Gauges

Ionization gauges are based on the measurement of ionization current that is dependent on gas density. This class of gauges includes

- Hot cathode ionization gauges
- Cold cathode ionization gauges

5.3.1 Hot Cathode Ionization Gauges

These gauges are commonly used in high and ultrahigh vacuum systems. They use a heated filament for the thermionic emission of electrons. The electrons are further accelerated by a grid to cause ionization of the residual gas. The positive ion current is measured and gives an indication of the gas density as shown in the relation

$$I_p = I_e \bar{l}\,\bar{\rho}\eta n \tag{5.7}$$

Since $P = nkT$

$$I_p = I_e \bar{l}\,\bar{\rho}\eta \frac{P}{kT} \tag{5.8}$$

where I_p is the ion current; I_e is the electron emission current; $\bar{l}$ is the average path length of electrons; $\bar{\rho}$ is the average electron-impact ionization cross section; η is the ion collection efficiency; n is the gas density; P is the pressure of the gas; K is the Boltzmann constant; T is the absolute temperature of gas.

The equation can be expressed in form

$$I_p = sI_e P \tag{5.9}$$

by maintaining the remaining terms constant. The average path length $\bar{l}$ of electrons is the average distance travelled by electrons till they are collected. $\bar{l}$ depends upon the electrostatic field of the gauge. The average electron-impact ionization cross section $\bar{\rho}$ depends upon the average electron energy. The sensitivity of the gauge called as s has different values depending upon the gas. For instance, for helium it is less by a factor of 10 as compared to N_2.

5.3.1.1 Schulz and Phelps Gauge

The Schulz and Phelps[6] ionization gauge has been designed for measurement of pressures beyond the upper measurement limit of the other hot cathode ionization gauges. Figure 5.7 shows the construction of the gauge.

The gauge design uses planar geometry. It contains two parallel plates of same size, spaced about 3 mm. One of the plates is the anode and the other is the ion collector. A filament is placed midway between them. The gauge offers a very short average path length of

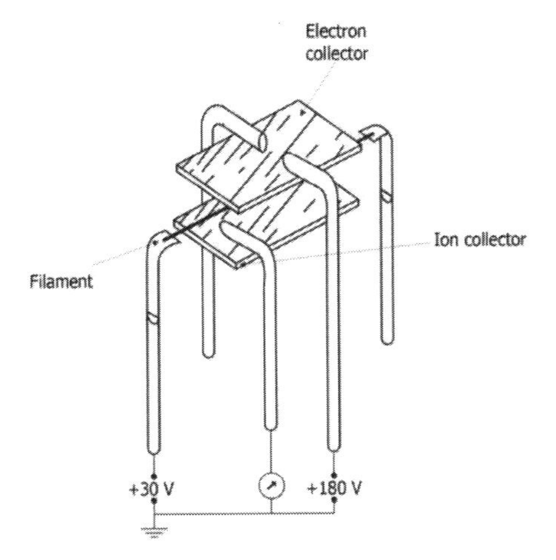

Fig. 5.7. Schulz and Phelps ion gauge.

electrons. Hence the gauge has low sensitivity and a higher pressure measurement limit of 100 Pa, The lower limit being 1×10^{-3} Pa.

5.3.1.2 Conventional Hot Cathode Ionization Gauge

Figure 5.8 shows the conventional ionization gauge. The axially located filament is heated by the passage of electric current and emits electrons. The electrons are accelerated by the cylindrical grid maintained at about 100–150 V positive with respect to the filament which is maintained at about 30 V positive with respect to the ion collector. The energized electrons cause ionization of the residual gas upon impact. The positive ions are collected by the ion collector while the electrons are collected by the grid. Typically emission currents of 10^{-3}A or 10^{-2}A are used. The range of pressure measurement with this gauge is 10^{-6} Pa to 10^{-3} Pa.

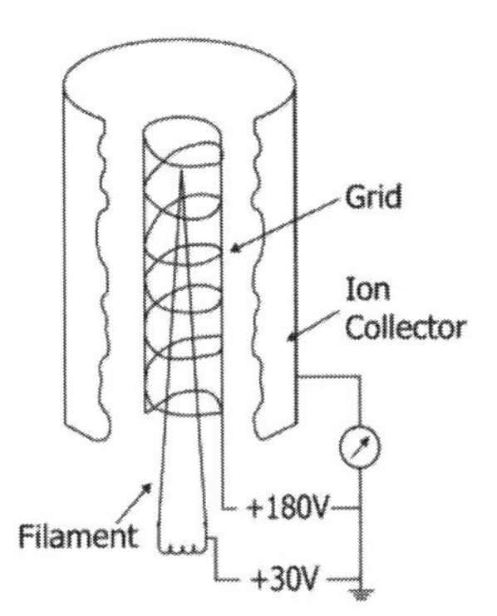

Fig. 5.8. Conventional ionization gauge.

Nottingham[7] established that the lower limit of 10^{-6} Pa in the hot cathode ionization gauge was due to the fact that the electron bombardment generates soft X-rays from the grid surface. The X-rays in turn release photoelectrons from the surface of the ion collector, thereby giving rise to a residual current, independent of pressure in the direction of the positive ion current.

In order to extend the low-pressure limit of the conventional hot-cathode ionization gauge it is necessary, at a given emission current, to increase the ratio of the ion current to the X-ray induced photo-current. This ratio may be increased by increasing the sensitivity of the ionization gauge. The gauge may be modified to permit the electrons to travel in long paths before they are collected by the positive grid or anode. Under these conditions the probability of the electrons colliding with an ionizing gas molecule will be enhanced, and the sensitivity of the gauge will be improved with no increase in X-ray photoemission.

5.3.1.3 Bayard–Alpert Gauge

The important observation of Nottingham paved way for the development of the Bayard–Alpert ionization gauge (B-A gauge) which has been a breakthrough in the measurement of low pressures.

Bayard and Alpert[8] interchanged the positions of the ion collector and the filament of the conventional ionization gauge. The filament is located outside the grid while the ion collector is in form of a fine wire suspended axially as shown in Fig. 5.9. With this arrangement, the solid angle subtended by the collector is several hundred time less than that in case of the conventional ionization gauge. The collector would intercept a small fraction of X-rays generated at the

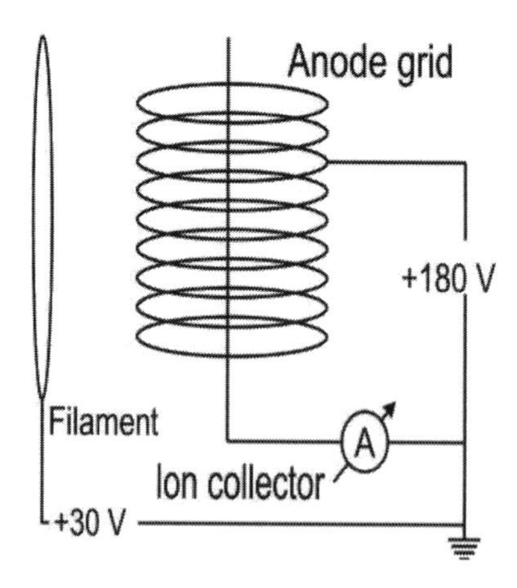

Fig. 5.9. Bayard–Alpert ionization gauge.

Fig. 5.10. Nude Bayard–Alpert gauge.

grid resulting in much lower emission of the photoelectrons, thereby lowering the pressure-independent residual current without any appreciable change in the sensitivity. This resulted in extending the limit of low-pressure measurement of the conventional hot cathode ion gauge by a factor of about 100.

Figure 5.10 shows photograph of a nude B-A gauge constructed by Naik and Verma.

Bayard and Alpert demonstrated this by measuring the ion current as a function of filament-to-grid voltage (electron energy) for a conventional ion gauge and a B-A gauge. At high pressures, both the gauges exhibited similar characteristics in agreement with the variation of the ionization cross-section with electron energy. However, for pressures of about 10^{-6} Pa the variation in the conventional gauge gave approximately a straight line with a slope of about 1.6 on the log–log plot which is a typical behavior of X-rays induced photoelectrons. At 4×10^{-7} Pa, the B-A gauge continued to show the gas ionization characteristics superimposed on a monotonically rising current due to photoelectrons. Finally, at about 5×10^{-9} Pa, the B-A gauge exhibited a characteristic similar to that for the conventional ion gauge. Thus, it was clear that in the B-A gauge the X-ray limit was reduced by a factor in excess of 100 and the pressures of the order of 10^{-8} Pa could be measured.

The reverse X-ray effect[9] occurs when a small fraction of the soft X-rays produced at the grid is intercepted by the ion collector at the ground potential and the remaining fraction is intercepted by the glass envelope with a conductive coating. If the photoelectrons emitted from the envelope possess enough energy to reach the ion collector, the resulting current at the ion collector is negative and is comparable in magnitude to that resulting from the forward X-ray effect caused by the photoelectrons originated at the grid. The reverse X ray effect can be eliminated if the glass envelope is biased 10 V positive with respect to the ion collector.

The logarithmic potential distribution between the grid and the collector of the B-A gauge with the maximum change in the potential immediately adjacent to the collector results to ionizing electrons retaining most of the energy within the grid volume. Thus effective ionization of the residual gas takes place.

The grid and the ion collector of the B-A gauge are 'degassed' for removal of the adsorbed gases on the metal surfaces to prevent erroneous measurements of pressure. This is achieved by electron bombardment of the grid and the collector by the electrons drawn from the filament with the grid and the collector electrically tied together at higher DC voltage. The degassing cleans the surfaces. However, the degassing results in an enhanced pumping action by the gauge. The pumping processes[10] associated with the gauge include

- Interaction of chemically active gases with the residual gases at the filament and at the envelop wall where the deposits are formed
- Ion pumping, predominantly of the inert gases that are ionized and are trapped at negatively charged electrodes and at the wall
- Removal of active neutral gases by attachment or combination at the wall resulting from excitation or disassociation of molecules within the gauge by electrons

It is therefore advisable to operate the gauge at low electron emission current of 1×10^{-4} A to minimize the pumping effects of the gauge. Accurate measurement of pressure is possible by considering the pumping speed of the gauge.

The range of pressure measurement of the B-A gauge is 1×10^{-8} Pa – 1×10^{-1} Pa.

Extension of the lower limit of the B-A gauge

Further developments in the hot cathode ion gauges were in the direction of minimizing the effects due to soft X-rays, surface generated ions and neutrals in order to extend the lower limit of measurement of pressure. Methods to achieve this extension include

- Measurement of residual currents and application of corrections to compute the true pressure as in the case of a modulated B-A gauge (to be discussed later)
- Minimization of X-rays reaching the ion collector by suitable positioning
- Suppression of emission of the photoelectrons from the ion collector
- Increasing the sensitivity of the gauge.

Nottingham[11] proposed employment of the additional grid-like enclosed ends for the grid and a screen grid in the form of an open cylinder surrounding the other components of the gauge. The end enclosure of the grid prevented the ions from escaping axially from the grid thereby improving the ion collection efficiency. The screen grid maintained at a high negative potential repels the electrons causing them to travel a longer effective path length and also shields the gauge from influence of charges on the glass envelop of the gauge.

Houston[12] developed an ionization gauge with the linearity of ion current with pressure at much lower pressures than the B-A gauge. The average path length of electrons of 10^8 cm was obtained by employing magnetic and electric fields to contain the ionizing electrons. The gauge provides high sensitivity and uses a lower filament emission current. Pressures of the order of 1×10^{-8} Pa were reported using this gauge.

5.3.1.4 Modulated Bayard–Alpert Gauge

Redhead[13] suggested ion current modulation by introducing an extra electrode into the grid space in the form of a wire close to the grid and parallel to the collector. With the wire at the grid potential, there is no significant effect on the gauge and the gauge would act as a normal B-A gauge. By lowering its potential by 100 V alters the ion trajectories, and the measured collector current is significantly reduced by the so-called modulation index without drastically perturbing the overall potential distribution within the grid. When measured at relatively higher pressure at about 10^{-4} Pa, the fraction *a* of the ions collected by the modulated electrode when operated at the lower potential is experimentally determined. At this pressure, the magnitude of the residual current due to the X-ray effect would be negligibly small compared to the pressure-dependent ion current. At an unknown low pressure, when the residual current I_s is a sizable fraction of the total ion collector current I_c, the true pressure-dependent ion current I_p can be determined as explained below.

$I_c = I_p + I_s$ .... when the modulator is at the grid potential, and
$I'_c = I'_p + I_s$ .... when the modulator is at the ion collector potential
Thus

$$I_c - I'_c = I_p - I'_p = \alpha I_p \tag{5.10}$$

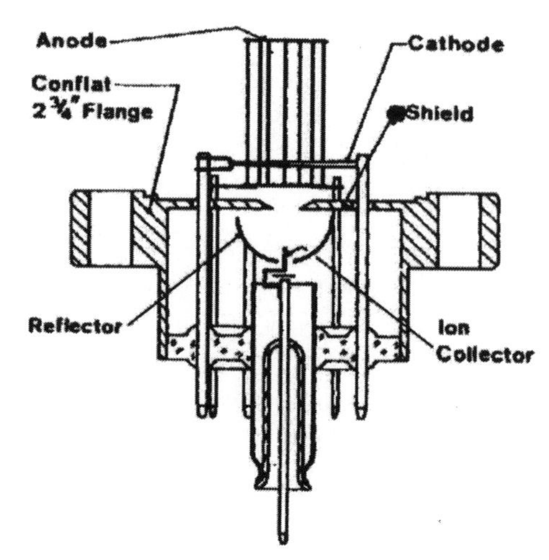

Fig. 5.11. Extractor gauge (Courtesy: CERN).

5.3.1.5 Extractor Gauge

Redhead[14] described this gauge in which the ion collector is removed away from the grid.

An electron emission current of 1.3 mA is used in this gauge. The ions are extracted from the ionization region, focused through an aperture in the shield electrode and collected on a short fine wire. The top of the grid is closed while the bottom is open. The ion reflector is in the form of the hemispherical electrode at the grid potential. The collector wire projects through a small hole in the ion reflector. The modulator is in the form of a short wire projecting into the grid on the opposite end of the ion collector as shown in Fig. 5.11.

The modulator potential is switched from the grid potential to the ground. This way, the ion current to the collector is modulated to about 50%. The X-ray flux exposed to the ion collector is reduced as the collector is partially shielded from the grid by the shield plate and as the collector is a short wire of small diameter. At pressure of 2×10^{-10} Pa the ion current equals the residual current due to X-rays. This results in an increase in the sensitivity. The lower limit of the pressure measurement is extended to about 10^{-11} Pa.

5.3.1.6 Magnetron Gauge

Lafferty[15] developed a magnetron gauge using a cylindrical anode and a magnetic field with an intensity of 2.5 times the cutoff value.

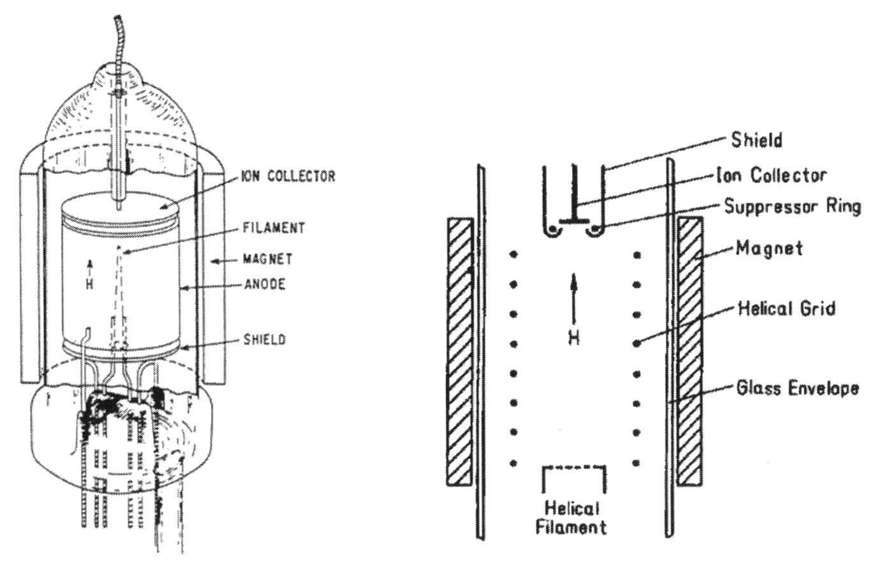

Fig 5.12 (left). Hot cathode magnetron gauge (*Reproduced with permission from "J. Appl. Phys. 32 (1961) 424.1961," – J.M. Lafferty, AIP Publishing LLC*).

Fig. 5.13: Hot cathode magnetron gauge with suppressor electrode (*Courtesy: CERN*).

Electrons emitted from the centrally located filament spiral around the cylindrical anode with large average path lengths. Two end plates maintained at a negative potential relative to the cathode to collect the ion current generated in the magnetron prevent the escape of electrons. The positive ion emission from the cathode is suppressed by mounting a hairpin filament on the axis of the cylindrical anode and well removed from the region of the negative ion collector. The electron emission current of 1 μA is used to prevent instabilities in operation, to minimize the ESD ions, to minimize the heating effects and to give a maximum ratio of ion current to X-ray photocurrent. Measurements of sensitivity and X-ray photocurrent indicate that the magnetron gauge is linear down to a pressure of 4×10^{-12} Pa. At pressure of 6×10^{-13} Pa, the ion current equals the residual current due to X-rays. The ion pumping speed was found to be 0.003 liters $\cdot$ s^{-1}.

The gauge is shown in Fig. 5.12. The lowest pressure of 1×10^{-14} Pa can be measured with this gauge[16] as its X-ray limit has been further reduced by the addition of a suppressor electrode in front of the collector as shown in Fig. 5.13.

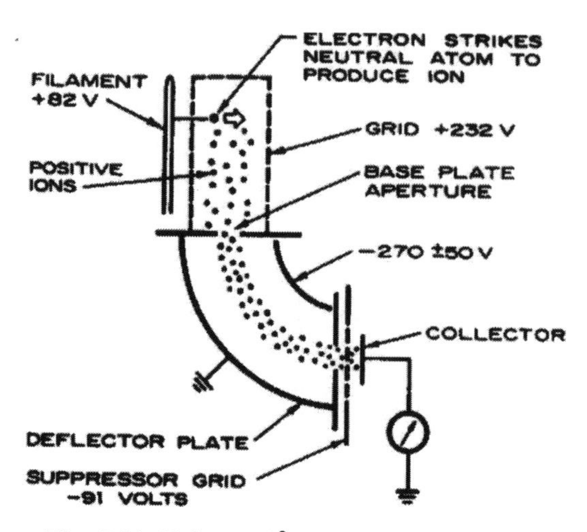

Fig. 5.14. Helmer 90° gauge (*Courtesy: CERN*).

5.3.1.7 Helmer Gauge

Figure 5.14 shows a schematic of the Helmer gauge. In the Helmer gauge[17] the filament is located outside the grid. The electron emission current of 3 mA is used in this gauge. The positive ions formed by electron impacts are extracted and deflected electrostatically by 90° before their collection onto the ion collector. With no line of sight between the grid and the collector, the X-ray limit is reduced to about 2×10^{-12} Pa which is the lowest measurable pressure for this gauge. The ions formed due to electron impact in the gas phase are separated from those formed due to (EID) in the gauge.

5.3.1.8 Helmer Gauge–180° Ion Deflection

Figure 5.15 shows the schematic of the 180° Helmer gauge. The electron emission current of 5 mA is used in this gauge which utilizes a hemispherical deflector. The ion collector plate is completely out of sight of the grid. A suppressor electrode inhibits electrons produced by reflected X-rays leaving the collector. With the inner electrode of the hemispherical deflector at the ground potential and the outer electrode at the variable positive potential, the ions generated can be separated based on their energy[18]. Ions formed at the grid resulting from (EID) possess higher energies compared to those formed in the gas phase resulting from the electron impact. This gauge can separate these ions more efficiently than the conventional Helmer gauge. A spherical grid is used to increase the electron space charge at its

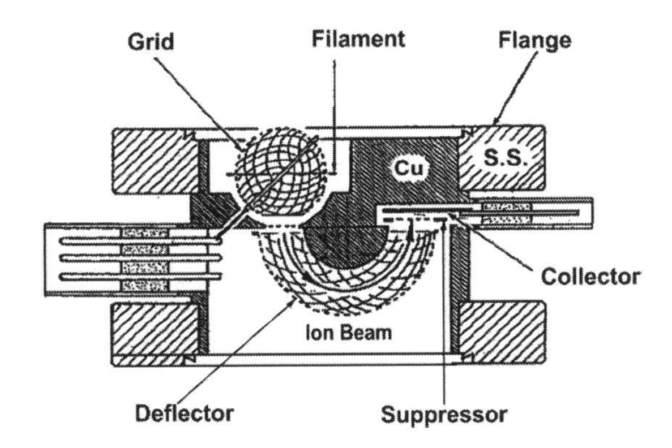

Fig. 5.15. Helmer gauge 180°–ion deflection–ion spectroscopy gauge (*Courtesy: CERN*).

centre. The pressure at which the ion current equals the residual current due to X-rays is <2.5 × 10^{-13} Pa. The lowest measurable pressure with this gauge is <10^{-12} Pa.

5.3.1.9 Axial Symmetric Transmission Gauge

The axial symmetric transmission gauge (A-T gauge) uses further modification[19] of the 180° Helmer gauge. It uses a 256.4° deflection of the ions. The electron emission current of 0.1 mA is used in this gauge. The pressure at which the ion current equals the residual current due to X-rays is <6 × 10^{-12} Pa. The lowest measurable pressure with this gauge is 4 × 10^{-13} Pa. The gauge is shown in Fig. 5.16

Figure 5.17 shows the schematic of Bessel-Box type energy filter for the A-T gauge. It has a secondary electron multiplier as the ion

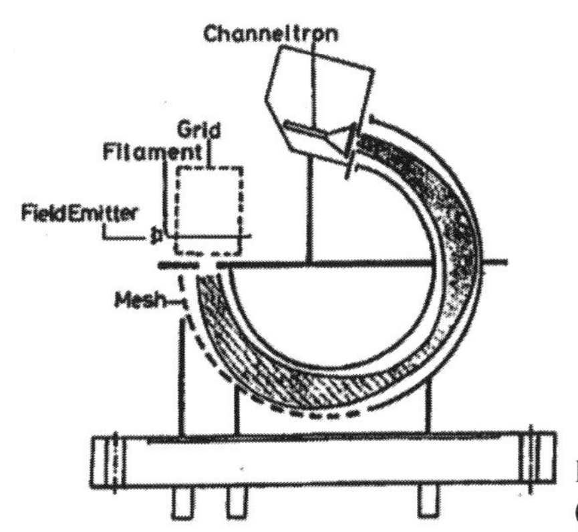

Fig. 5.16. 256.4° Beam deflection gauge (*Courtesy: CERN*).

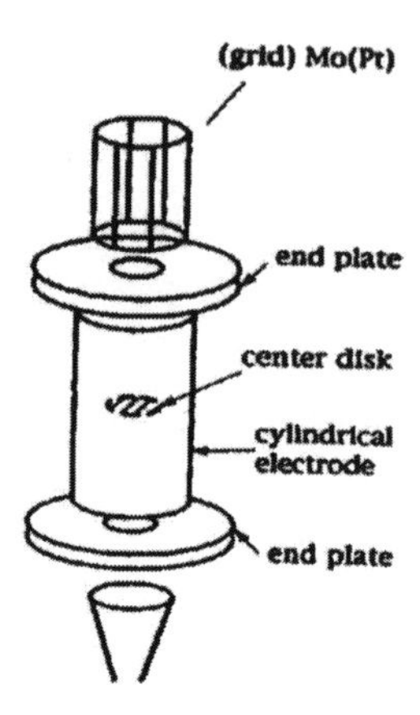

Fig. 5.17. Bessel-Box type energy filter (*Courtesy: CERN*).

collector. The Bessel–Box type energy analyzer has a cylindrical electrode with a centre disc and two end plates. The centre disc, placed perpendicularly to the cylinder axis, shields the ion collector from direct incidence of X-rays generated at the surface of the ionizer. Ions with less energy are retarded and collected while the ions with excessive energy will not pass through the exit aperture. The energy analyzer focuses the ions from the limited region in the grid cage on the ion collector. Akimichi et al[20] have observed that the energy spectrum obtained by the energy analyzer showed two peaks of gas-phase ions and electron-stimulated desorption (ESD) ions, completely separated with an energy difference of 30 eV.

The residual X-ray photocurrent was equivalent to a pressure 3.5×10^{-11} Pa (N_2 eq.). They used an emission current of 30 μA. The lower limit of pressure measurement of the gauge was estimated to be 3×10^{-12} Pa or less.

5.3.1.10 Orbitron Gauge

The orbitron gauge[21] is based on the radial electrostatic field between two concentric cylinders. The central cylinder is in form of a thin wire or a metallized quartz rod. The outer cylinder is the glass envelope of the gauge that has an internal metallic electrically conductive coating and serves as the ion collector. A hot filament is located in the electric field as shown in Fig. 5.18. The filament is biased at a positive potential which is less than the electrostatic potential at its position in the field. A short length wire is mounted close to the filament between the filament and the anode. The electrons from the filament injected in the field possess angular

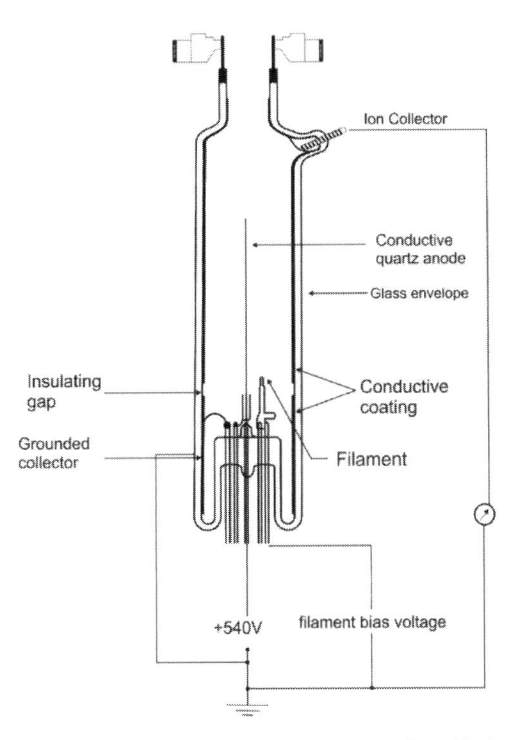

Fig. 5.18. Orbitron gauge (*Reprinted from J. Vac. Sci. Technol., 4, 63 (1967), E.A. Meyer and R.G. Herb with the permission of AIP Publishing LLC*).

momenta which have a component also in the axial direction. The injected electrons move into orbits around the central anode. Electron orbits for varying angular momenta are described by Hooverman[22]. The electrons have a large average path length of several thousands of centimeters. A lead passing through the external envelop of the gauge measures the ion current which is proportional to pressure. Only 1 μA electron current is required to produce the same ion current obtained from an 8 mA emission in the B-A gauge. The total power input to the gauge is less than 100 mW. The ion current is approximately linear with pressure from 10^{-9} Pa to 10^{-2} Pa. The linear region can be extended to 1.0 Pa by operation in a non-orbiting mode.

Limitations and advantages of hot cathode ion gauges

Limitations:

- Variation of sensitivity with different gases. This is primarily due to different average ionization cross section values for the gases. Commercially available gauges generally specify the sensitivity constant for N_2. To calculate the true pressure of a gas, one needs to multiply the N_2 equivalent pressure by a factor of the ratio of sensitivity for N_2 to the sensitivity of the particular gas

- Chemical reactions occurring on the hot filament surface change the gas composition
- Pumping effect is caused by the gauges
- Non-linearity of ion current versus pressure exists at low pressures due to X-ray induced current
- Thermal and electron-induced gas desorption occurs from anode
- Formation of surface generated ions and neutrals

Advantages:

- The gauges cover a wide range of pressure measurement

5.3.2 Cold Cathode Ionization Gauges

5.3.2.1 Penning gauge

Also known as the Philips ionization gauge, the Penning gauge is shown in Fig. 5.19. The gauge consists of a central anode in the form of a circular ring or a short length cylinder located midway between a pair of flat circular cathodes. The anode is maintained at about 2 kV dc. An axial magnetic field is applied. The discharge current is measured and is proportional to the gas pressure in the range 10^{-6} Pa and 1 Pa which is the useful range of pressure measurement of the Penning gauge. The electric field and the magnetic field are effectively crossed due to a negative space charge formation along the axis at the centre of the anode. The average path length of the electrons is sufficiently long. The advantages include a robust and simple construction and absence of the hot filament. The limitations include the non-linearity of the discharge current with pressure, the sensitivity variation with gas, presence of the magnet, ion pumping and difficulty in starting at low pressures.

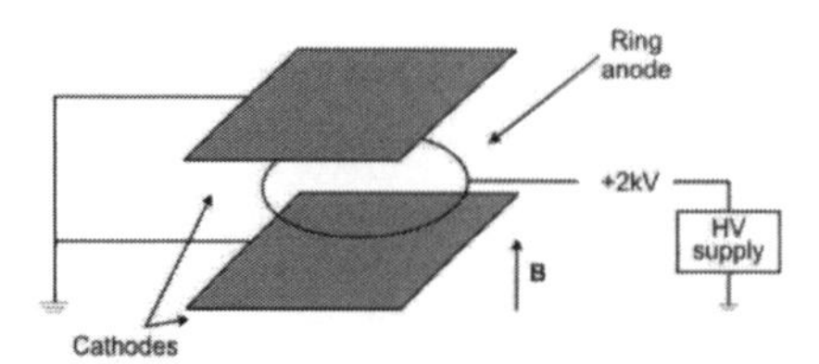

Fig. 5.19. Penning gauge.

5.3.2.2 Magnetron and Inverted Magnetron Gauges

Hobson and Redhead[23] developed the magnetron and inverted magnetron gauges, the latter was earlier designed by Haefer[24] in 1955.

Figure 5.20 shows the schematic of the cold cathode magnetron gauge. The anode is in the form of an open cylinder of 30 mm diameter and 20 mm length. The anode is perforated to improve the gas flow through the gauge. The spool shaped cathode consists of axial cylinder of 3 mm diameter and 20 mm length. The cathode is welded to two circular end discs which are shielded by two annular shield auxiliary electrodes operated at the cathode potential. These electrodes are interposed between the anode and the end plates to minimize field emission. A conductive film deposited on the glass envelop is electrically connected to the auxiliary electrode. The film acts as an electrostatic shield.

The gauge is normally operated with an axial magnetic field of 1000 Gauss and the anode voltage of 4.5 to 6 kV.

The inverted magnetron gauge[25] consists of two co-axial cylinders. The inner cylinder is the anode rod of 1 mm diameter and the outer cylinder is the ion collector of 30 mm diameter and 20 mm length. The ion collector is partially closed at both ends with its axis parallel to the magnetic field. The auxiliary cathode is a box surrounding the ion collector to which are welded two short tubes which extend inside the end plates of the ion collector. The gauge is shown in Fig. 5.21. The auxiliary cathode acts as an electrostatic shield and also prevents field emission from the edges of the hole in the ion collector end plates. The ion collector and the auxiliary cathode are maintained at the ground potential. The gauge can be operated at up to 6 kV with 0.2 Tesla. The ion current I_p was found to follow a relation given by

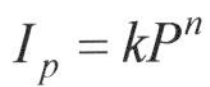

$$I_p = kP^n$$

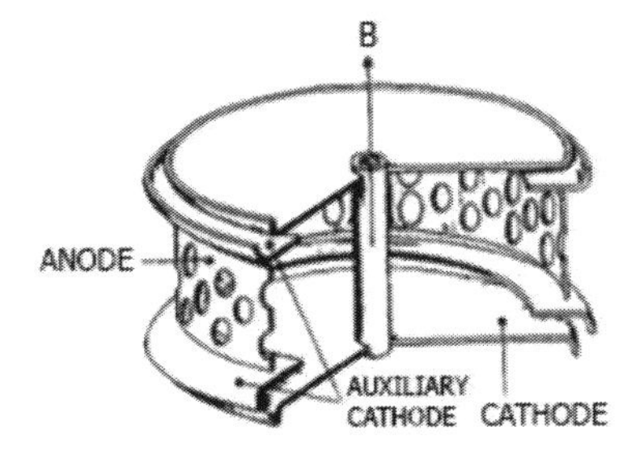

Fig. 5.20. Cold cathode magnetron gauge (*P.A. Redhead – The Magnetron Gauge: A Cold-Cathode Vacuum Gauge – Canadian Journal of Physics, 1959, 37(11): 1260-1271, 10.1139/p 59-144 – © Canadian Science Publishing or its licensors*).

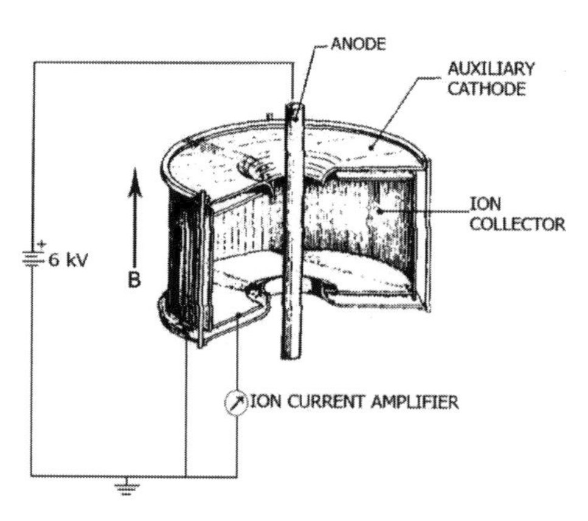

Fig. 5.21. Cold cathode inverted magnetron gauge (*J.P. Hobson, P.A. Redhead – Operation of an Inverted-Magnetron-Gauge in the Pressure Range 10^{-3} to 10^{-12} mm Hg, Canadian Journal of Physics, 1958, 36(3): 271-288, 10.1139/p 58–031 – ©Canadian Science Publishing or its licensors*).

where P is the pressure and k is a constant ≈1.10 to 1.15.

The gauge efficiently traps the electrons. The discharge current is stable to much lower pressures. Hobson and Redhead[23] claimed that the gauge can be used for pressure measurements in the range 10^{-11} Pa – 10^{-2} Pa.

Cold cathode gauges are rugged and cannot 'burn out' due to the absence of the filament. These gauges have no X-ray limit to the lowest measurable pressure. Cold cathode gauges can be contaminated if operated at pressures higher than 10 Pa. The presence of argon causes sputtering of the cathode. The getttering action on the surfaces of the cathode results in pumping that can cause errors in the measurement of pressure. These gauges have difficulty in striking at pressures less than about 1×10^{-7} Pa.

Watanabe[26] has reviewed the total pressure measurement gauges in the extreme high vacuum region and has discussed development of the point collector gauge and the modulated ion current gauge. He claims to have overcome the major barriers in the extreme high vacuum pressure measurements that include the X-ray limit and the electron-simulated desorption of ions. He has further described the use of a copper housing and a heated grid gauge to counter the outgassing electron-stimulated desorption of neutrals.

5.4 Partial Pressure Gauges

Knowledge of partial pressures and their contribution to the total pressure is important in research and technology. Partial pressure gauges (PPGs) provide details of the quality of vacuum as they reveal the composition of the residual gases in vacuum. All PPGs have three features in common. These include:

- An ion source and ion acceleration in a given direction
- An analyzer to sort out ions based on e/m ratio using electric and/or magnetic fields
- A detector where the ions are collected

Important factors that should be considered for employing PPGs, particularly at very low pressures, include:

- Sensitivity
- Resolving power
- Outgassing rate

If the partial pressure of the order of 10^{-11} Pa is to be detected, the sensitivity of the PPG should not be less than 10^{-6} A/Pa so that the measurement of the ion currents is limited to 10^{-17}A. It should be possible to have distinct separation of the residual or the generated gas in the system with the available resolving power of the PPG. Generally, a resolving power of 50 at mass number 44 (CO_2) is considered adequate for most partial pressure measurements. The PPG should be bakeable to 400° to 450°C and its outgassing rate must be less than about 10^{-9} Pa · liters/s during the normal operation for UHV applications.

5.4.1 Magnetic Sector Mass Spectrometer

The magnetic sector-type mass spectrometer was designed by Alfred Nier[27].

The positive ions generated in the ion source by the electrons are accelerated and the mono-energetic beam of ions is injected into the magnetic field through a narrow slit in the base of the ionization chamber. The ions are bent in circular paths through 180° towards the slit in the collector plate. The radius of curvature of path r is decided by

- Strength of the magnetic field
- Charge to mass ratio of the ions and
- Energy of the ions.

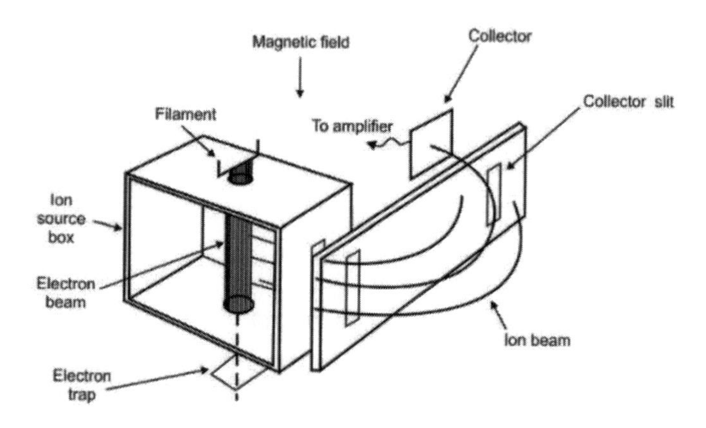

Fig. 5.22. 180° deflection mass spectrometer.

Figure 5.22 illustrates the concept of the 180° deflection mass spectrometer.

The kinetic energy of the ions is given by

$$\frac{mv^2}{2} = eV \tag{5.11}$$

$$v = \left(\frac{2eV}{m}\right)^{\frac{1}{2}}$$

V is the voltage across which the ions are accelerated in the ion source.

If B is magnitude of the magnetic field, m is the mass, e is the charge and v the velocity of the ion in the magnetic field then r is the radius of curvature of the ions is given by

$$\frac{mv^2}{r} = Bev \tag{5.12}$$

$$\left(\frac{m}{e}\right) = \frac{B^2 r^2}{2V} \tag{5.13}$$

Only those ions having a radius exactly equal to r reach the collector plate through the slit. The heavier and lighter ions with different radii cannot reach the collector. By controlling the accelerating voltage V, the kinetic energy of the ions can be changed so that components of the ion beam can be scanned.

Fig. 5.23. Assembly of the nude mass spectrometer.

The resolving power R of the magnetic sector type is given by

$$R = \frac{m}{\Delta m} \tag{5.14}$$

where m is the mass number and Δm is the difference between two masses that can be separated. In the case of a single peak, the resolving power R = 5% of the peak width/mass of the peak. Also, if each mass can be separated from the next mass, then it amounts to unit resolution.

Figure 5.23 shows a photograph of the flange mounted assembly of the nude type magnetic sector, 1 cm radius–180° deflection mass spectrometer constructed and used by Naik[28].

5.4.2 Omegatron

Alpert and Buritz[29] have described the omegatron. Figure 5.24 shows the schematic diagram of the omegatron. The subassembly of omegatron constructed at Herb's laboratory at the University of

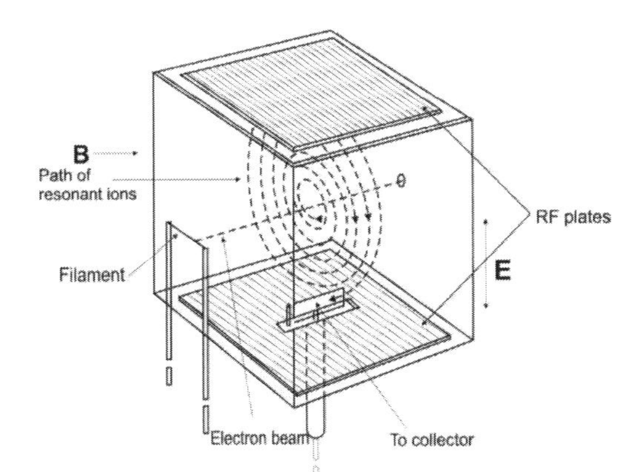

Fig. 5.24. Schematic diagram of the omegatron.

Fig. 5.25. Omegatron sub-assembly.

Wisconsin, USA is shown in Fig. 5.25. An RF field is applied across the parallel insulated electrode plates of the omegatron. A constant magnetic field H is crossed with the RF field. A fine electron beam causes ionization of the residual gas. Positive ions formed are made to travel in spiraling paths by the axial magnetic field and the RF field. The time of revolution T in the magnetic field B is given by

$$T = \frac{2\pi m}{eH}$$

The time T is depends upon the strength of the magnetic field and the mass of the ions irrespective of energy. The ions of a particular mass m and charge e resonate with the RF field applied and gain energy continuously. The radius of curvature of their paths increases until the spiraling ions are incident on the collector. The non-resonant ions do not gain enough energy to reach the collector. The ion current at the collector is proportional to the gas density of the ion species. The mass spectrum can be obtained by varying either the strength of the magnetic field or the RF frequency. Resolution is inversely proportional to mass in the omegatron. Thus its performance is best suited for lighter ions in the range 1–4 amu.

5.4.3 Quadrupole Mass Analyzer

In the quadrupole, mono-energetic positive ions are extracted from an ion source. The ions of different masses are further injected into

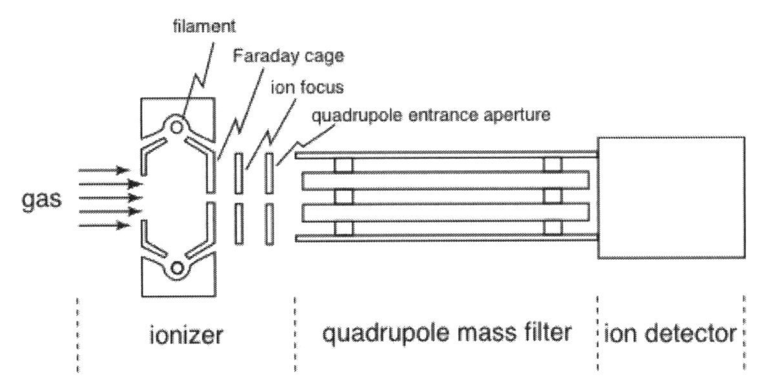

Fig. 5.26. Operating principle of a quadrupole mass spectrometer (Courtesy: Philip Hofmann).

the axis of the electrostatic mass analyzer assembly that consists of four parallel metal rods arranged as shown in Fig. 5.25.

Although ideally the quadrupole fields require rods with a hyperbolic profile, in practice, rods with the circular cross section are used. An electrical quadrupole field is formed between the rods. One pair of opposite rods is held at a potential of $[U + V\cos(\omega t)]$ and the other pair is held at a potential $-[U + V\cos(\omega t)]$, where U is DC voltage and $[V\cos(\omega t)]$ is the AC voltage. The applied voltages influence the trajectory of ions traveling down the flight along the axis of the rod assembly. For given DC and AC voltages, only those ions of a certain mass-to-charge ratio pass through the quadrupole filter and are collected at the ion collector. All other ions are deviated away from the original path. A mass spectrum is obtained by monitoring the ions passing through the quadrupole filter as the voltages on the rods are varied. This can be achieved by varying ω while maintaining U and V constant, or varying U and V (U/V) with constant ω.

5.4.4 Time-of-Flight (TOF) Mass Analyzer

Stephens[30] proposed the concept of the TOF analyzer. This analyzer is based on the principle that ions with the same kinetic energy but different masses travel with different velocities. The ions are separated on the basis of differences in their velocities. Ions formed by a short ionization event are accelerated by an electrostatic field to a common energy. These ions further travel over a field-free drift path to the detector. The lighter ions arrive before the heavier ions and a mass spectrum is recorded. Measuring the flight time for each

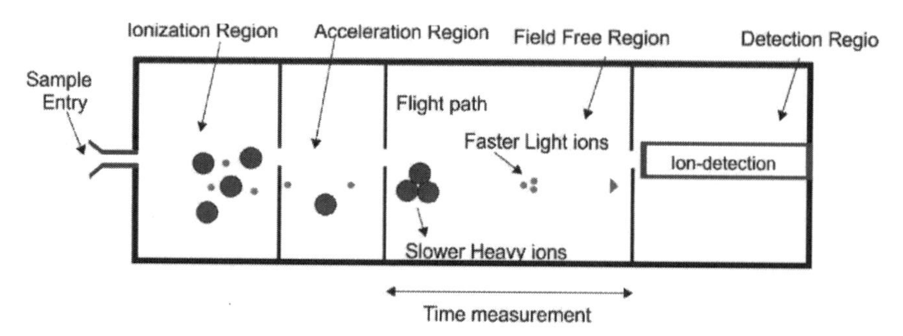

Fig. 5.27. Schematic of time-of flight mass spectrometer.

ion allows the determination of its mass. Figure 5.27 shows a TOF analyzer.

Ions of the same charges, as they emerge from the ionization chamber, possess equal kinetic energies. However, their velocities depend upon the masses of the ions. The ions drift in the field-free flight tube with these velocities. The kinetic energy of the ions is given by

$$\frac{mv^2}{2} = eV$$

where V is the accelerating voltage for electrons in the ionization chamber that causes ionization; m, v and e are the mass, velocity and the electric charge respectively of the ion.

The time of flight, or time it takes for the ion to travel the length L of the flight tube is:

$$t = \frac{L}{v}$$

Substituting the equation for kinetic energy in the equation for the time of flight, we have

$$t = L\left(\frac{m}{2eV}\right)^{\frac{1}{2}} \tag{5.15}$$

During the analysis, L, the length of tube, V, the voltage from the ion source, are all held constant. Thus, the time-of-flight is directly proportional to the root of the mass to charge ratio. At higher masses, the flight time is longer and not all the ions of the same m/e values attain their ideal TOF velocities. A reflectron consisting of a series of ring electrodes of high voltage is placed at the end of the drift tube.

When a heavy ion travels into the reflectron, it is reflected in the opposite direction. The reflectron increases resolution by narrowing the broadband range of flight times for a single *m*/*e* value. Both slow and fast ions, of the same *m*/*e* value, reach the detector at the same time rather than at different times, narrowing the bandwidth for the output signal. The TOF analyzer has high resolving power and accuracy. It is particularly useful for the determination of large bio-molecules and for chromatography.

5.4.5 Detectors

The ions that are separated on the basis of their mass-to-charge ratio can be detected and monitored using different types of detectors such as:

- Faraday cup with an electrometer amplifier for direct measurement of the ion current
- Secondary electron multiplier with individual dynodes
- Continuous secondary electron multiplier

Detector selection is governed by the requirements of detection sensitivity, detection speed and signal-to-noise ratio in addition to stability, thermal and chemical resistance.

5.4.6 General Discussion on Residual Gas Analyzers

Terms, commonly used in mass spectrometry are discussed.

Scanning: The variation of the parameters such as magnetic field or ion accelerating voltage in the case of magnetic sector type mass spectrometers and the variation of RF and DC voltages in quadrupole analyzers that causes the ions of different *m*/*e* ratios to be collected at the detector is called scanning.

Mass spectrum: The mass spectrum is a plot representing the distribution of detected ions. The masses of ions (the masses divided by their charge) appear on the *x* axis. The charge on the ion is assumed as 1+. Figure 5.28 shows a typical mass spectrum indicating the variation of ion current at the collector resulting from scanning. Ions of a particular *m*/*e* ratio correspond to the value of the parameter varied during the scanning. Thus the peaks in the spectrum can be identified with the particular mass numbers. The magnitudes of the partial pressures correspond to the heights of the individual peaks. The abundance is expressed

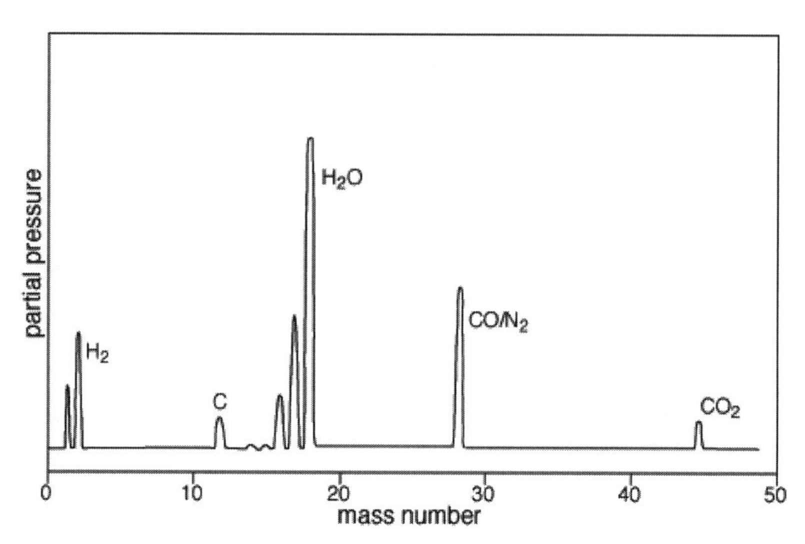

Fig. 5.28. Typical mass spectrum (*Courtesy: Philip Hofmann*).

either in an absolute or in a relative form. The most intense abundance is considered as 100 % and the other intensities are normalized to this value. The heaviest ion (the one with the greatest m/e value) is likely to be the molecular ion in the mass spectrum.

Sensitivity: The value of the ion current for the dominant peak of gas sample when the pressure at the source is 1 Pa and is expressed in A/Pa. Generally RGAs offer sensitivity in the range of 10^{-5} to 10^{-6} A/Pa.

Resolution: The ability to resolve the adjacent mass peaks is called the resolution. The two neighbouring equal peaks are considered as resolved if the valley between them is 10% of either from the base line. Generally, the resolving power is represented by the mass number up to which valley of 10% or less can be obtained.

Cracking Pattern: The table of relative intensities of the fragment ions formed in the RGA considering the intensity of the most dominant peak as 100% is shown in the cracking pattern. Disassociation of gas molecules takes place resulting from the electron impacts in the ion source. Few compounds have mass spectra without any molecular ion peak, as all the molecular ions break into fragments.

Interpretation of Mass Spectrum: Complications can arise in interpretation of the mass spectrum resulting from the possibility of existence of different isotopes in the molecular ion. The presence of the carbon-13 isotope in a molecular ion may cause another adjacent small peak. Similarly, a compound containing chlorine atom may cause two adjacent small peaks due to the presence of isotopes chlorine-35 and chlorine-37.

In most cases, one considers *m*/*e* values in a mass spectrum as whole numbers as the rounded-off values of isotopic masses are used. By using a high resolution mass spectrometer, it is possible to obtain far more accurate results by using more accurate information of the masses to 4 decimal places such as

^{1}H	1.0078
^{12}C	12.0000
^{14}N	14.0031
^{16}O	15.9949

The carbon value is considered 12.0000, as all the other masses are measured on the carbon-12 scale which is based on the carbon-12 isotope having a mass of exactly 12.

A typical mass spectrum is shown in Fig. 5.28.

The main residual gases in this vacuum system are hydrogen, water, carbon monoxide and some carbon dioxide. It is important to note that the criterion for some test gas molecules to pass the mass spectrometer is the right *e*/*m* and not simply *m*. This means that doubly-charged ions appear as particles with half the mass in the spectrum. CO does for example not only give a peak at $m = 28$ but also one at $m = 14$ due to double ionization. The gases are also dissociated in the spectrometer such that one does not only find one peak for single-ionized water at $m = 18$ but also peaks at $m = 16$, 17 and 2 for the fragments.

Without any bakeout, the pressure of the system will be dominated by a high partial pressure of water. After the bakeout of the system, this will be significantly reduced and the total pressure will be determined by CO and hydrogen. If there is an air leak in the system this would show up as peaks of 28 (N_2) and 32 (O_2). The peak at 28 is always present due to CO and a small peak of 32 might even be present in a leak-tight chamber. However, more conclusions can be drawn from the cracked fragments if the peak at 14 (N) is bigger than the peak at 12 (C), this is usually an indication of an air leak.

References

1. Lord Rayleigh, *Phil. Trans. R. Soc. Lond. A*. **196**, 205 (1901).
2. D. Alpert, C. G. Matland, and A. O. McCoubrey, Rev. Sci. Instr. **20**, 370 (1951).
3. M. von Pirani, Deutsche Physikalische Gesellschaft, Verh. **8**: 24 (1906).

4. N. R. Campbell, Proc. Phys. Soc. London **33**, 287 (1921).
5. S. J. Chang, T. J. Hsueh, C. L. Hsu, Y. R. Lin, I. C. Chen and B. R. Huang, Nanotechnology, **19** (9) (2008).
6. G. J. Schulz and A. V. Phelps, Rev. Sci. Instrum. **28**, 1051 (1957).
7. W. B. Nottingham, Conf. Phys. Electronics (1947).
8. R. T. Bayard and D. Alpert, Rev. Sci. Instr. **21**, 571 (1950).
9. A. Berman, "Total Pressure Measurements in Vacuum Technology", 1st Edition, Academic Press (1985).
10. D. Alpert, Handbuch der Physik, **XII**, Ed: S. Fluggee/Marburg, Springer Verlag, Berlin (1958).
11. W. B. Nottingham, 1954 Vac. Symp. Tran. P. 76, Boston: Committee on Vacuum Techniques.
12. J. M. Houston, Bull. Amer. Phys. Soc. (II) **1**, 301 (1956).
13. P. A. Redhead, Rev. Sci. Instrum. **31**, 343 (1960).
14. P.A. Redhead, J. Vac. Sci. Technol. **3**, 173–180 (1966).
15. J. M. Lafferty, J. Appl. Phys. **32**, 424 (1961).
16. J. Z. Chen, C.D. Suen and Y.H. Kuo, J. Vac. Sci. Technol. A. **5**, 2373 (1987).
17. J. C. Helmer and W.H. Hayward, Rev. Sci. Instrum. **37**, 1652 (1966).
18. F. Watanabe, J. Vac. Sci. Technol. A **11**, 1620 (1993).
19. C. Oshima and A. Otaku, J. Vac. Sci. Technol. A **12**, 3233 (1994).
20. K. Akimichi, T. Tanaka, K. Takeuchi , Y. Tuzi and I. Arakawa, Vacuum **46**, (8–10), 749 (1995).
21. E. A. Meyer and R. G. Herb, J. Vac. Sci. Technol. **4**, 63 (1967).
22. R. H. Hooverman, J. Appl. Phys. **34**, 3505 (1963).
23. J. P. Hobson and P.A. Redhead, Can. J. Phys. **36**, 271 (1958).
24. R. Haefer, Acta Physica Austriaca, **7**, 52 (1953); **7**, 251 (1953) and **8**, 213 (1954).
25. P. A. Redhead, Can. J. Phys. **37**, 1260 (1959).
26. F. Watanabe, Vacuum **53**, (1–2), 1 (1999).
27. W. Johnson and A. Nier, Phys. Rev. **105** (3), P. 1014 (1957).
28. P. K. Naik, "Physical Processes in Vacuum – The Study of Entrapment of Inert Gas Ions into Polycrystalline Molybdenum", Ph.D. Thesis, Bombay University (1979).
29. D. Alpert and R.S. Buritz, J. Appl. Phys. **25**, 202 (1954).
30. W. E. Stephens, Phys. Rev. **69**, 691 (1946).

6

Vacuum Pumps

Pumps for generating vacuum are mainly classified in the following categories.

Positive displacement pumps

In these pumps, the gas to be pumped is trapped, compressed and ejected. Water-ring pumps and all mechanical rotary pumps including oil-sealed pumps, Roots type pumps and turbo-molecular pumps fall in this category. In the other positive displacement pumps such as steam/vapour ejector pumps, diffusion pumps, displacement of the gas is caused by the action of vapour streams on the gas to be pumped.

Pumps using conversion of gas into solid phase

This category of pumps includes getter pumps, getter-ion pumps, sorption pumps, cryogenic pumps, cryo-sorption pumps. In these pumps, the gas to be pumped is converted into solid phase within the pump and is not ejected.

6.1 Positive Displacement Pumps

6.1.1 Water-Ring Pump

The water-ring pump shown in Fig. 6.1 is a rotary positive displacement pump. The impeller forces the water by centrifugal acceleration to form a moving cylindrical ring on the interior of the casing. The water ring creates a series of seals in the space between the impeller vanes. The area of void space without water, sealed off between the water ring and the impeller blades is termed as 'impeller cell'. As each cell rotates past the inlet port, it carries a volume

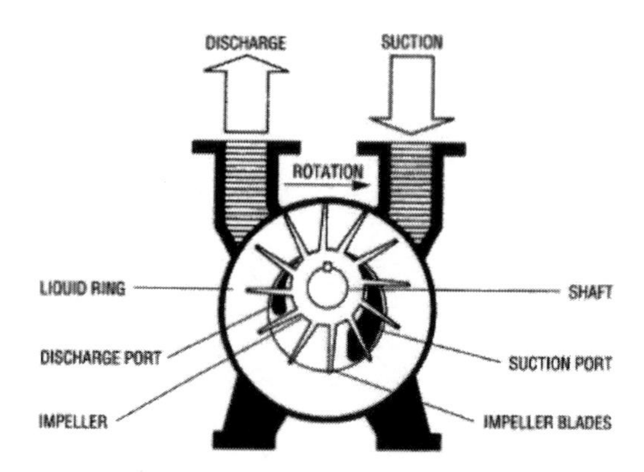

Fig. 6.1. Schematic of water-ring pump *(Courtesy: Sugartech).*

of air around with it. The gas/air is compressed as the liquid ring converges with the cone due to reduction in its volume. The eccentric mounting of the impeller in the casing causes cyclic variation of the volume trapped by the vanes and the ring. When each cell reaches the discharge port opening, the compressed gas escapes from that cell through the discharge port to the internal discharge passage.

These pumps find applications in pharmaceutical, glass, chemical industry and in oil exploration and refining.

6.1.2 Rotary Vane Pump and Rotary Piston Pump

Rotary vane and piston pumps are mechanical pumps that can discharge the pumped gas directly against the atmospheric pressure. It is important to note that the actual pumping speed S of mechanical rotary pumps can never drop to zero because the volumetric rate of displacement always remains unchanged as long as the rotational speed (RPM) of the rotor is constant. The effective speed, however can reduce and will be zero as one does not observe any fall in pressure at the limiting pressure.

Earlier, we have seen from equation 2.35 that

$$P_u = \frac{Q}{S} \quad \text{when} \quad \frac{dP}{dt} = 0$$

And

$$S = \frac{Q}{P_u}$$

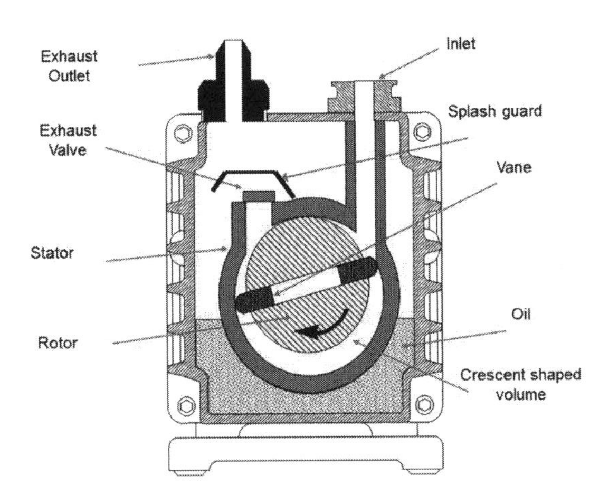

Fig. 6.2. Rotary vane pump *(Courtesy: Precision Plus)*.

This indicates that at the ultimate pressure P_u, the pumping speed S is handling the gas influx rate Q which is due to source of gases other than the intentional gas leak. In case of mechanical pumps, Q is mainly due to return of the pumped gases from the compression stage to the inlet port of the pump and due to the vapour pressure of the pump oil.

A typical rotary vane pump is shown in Fig. 6.2. The rotor is mounted eccentrically in the body of the stator. A pair of vanes in the form of rectangular plates is fitted diametrically opposite using springs on the rotor. The vanes have curved contours to suit the curvature of the inside surface of the stator. The contoured surfaces of the vanes press against the stator wall by springs. The layer of oil between the vanes and the stator causes the vacuum seal.

The system to be evacuated is connected to the inlet of the pump. As the pump is started (Figs. 6.3 *a* and *b*), the pressures in the trapped region and in the system are equal. On further rotation, the volume of the entrapped gas reduces, thereby causing compression at higher pressure (Fig. 6.3 *c*). As the rotor continues to move, the pressure exceeds the sum total of the pressures including the atmospheric pressure, the oil column pressure and the outlet valve spring pressure. Thus, the gas is ejected through the oil and the outlet valve (Fig. 6.3 *d*).

The oil that surrounds the stator serves to

- seal the joints from leaks
- lubricate the moving parts
- cool the pump assembly

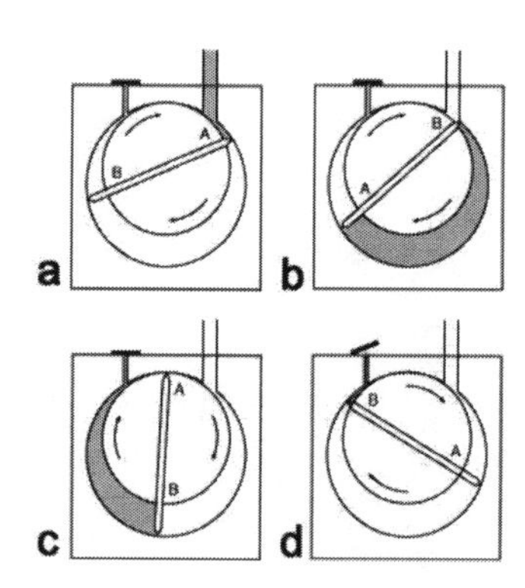

Fig. 6.3. Working of rotary vane pump (*Courtesy: Philip Hofmann*).

Oil ways on the external surface carry the oil into the interior of the stator where a film of the oil is formed. The oil is required to be chemically stable, have low vapour pressure and viscous enough to form leak-free seals between the stator and the rotor, rotary vanes.

The gas ballast method is used to pump condensable vapours such as water vapour that might be present in large quantity in the system. In absence of gas ballast, the low-pressure gas from the inlet may reach a pressure at which the water vapour can condense. The condensed vapour can mix with the pump oil that could give rise to a high vapour partial pressure in excess of about 4300 Pa at 30°C. The liquid dilutes the oil, and may cause corrosion of the pump body. The gas ballast valve, introduces a small amount of dry gas such as nitrogen. This helps in reducing the compression that the vapour can undergo, thereby reducing the condensation. The limiting pressure of the pump is adversely affected by using gas ballast. The oil must be chemically stable.

The useful operating range of the pump is between 10^5 Pa at the higher end and 1–5 Pa at the lower end. Ultimate pressures of the order of 2×10^{-1} Pa can be achieved using these pumps. Figure 6.4 shows variation of the effective pumping speeds of two typical rotary vane pumps of different capacities with inlet pressure. At lower pressures the effective speed falls as the gas influx from the sources other than the gas in the volume of the system dominates though the volumetric speeds remains unchanged.

The low-pressure limit on this pump is governed by

- vapour pressure of the oil
- outgassing
- leakage of gas around the seals

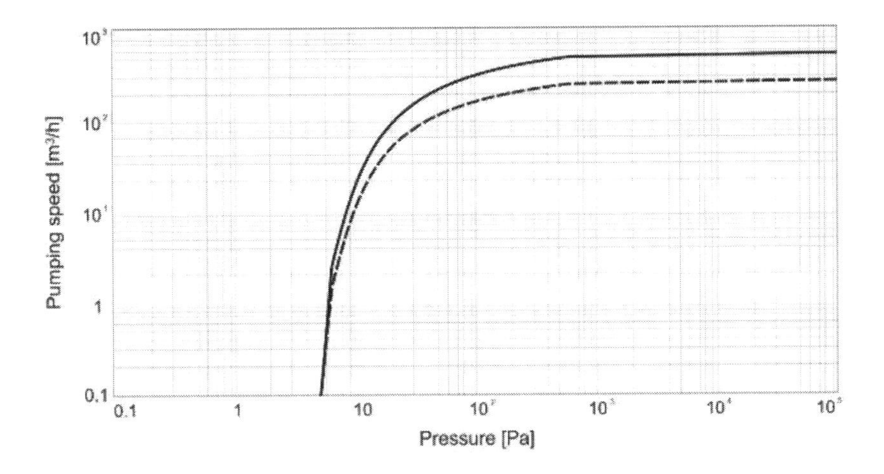

Fig. 6.4. Variation of pumping speed for air with pressure of typical rotary pumps.

During pumping, the gas from the compressed region of the pump can leak back through the oil seal between the rotor vane and the stator. Also, in the molecular flow pressure region, the oil molecules of vapour from the roughing pump can backstream into the system, thereby causing undesirable contamination of the system. A cold trap can be employed to minimize backstreaming. To prevent the undesirable entry of the pump oil from the stator into the roughing line, a float valve is generally employed that isolates the roughing line from the stator. In case the roughing line is to be exposed to the atmospheric air upon switching off the pump, the atmospheric air is introduced though a trap containing P_2O_5 to prevent moisture in the air entering the pump and the roughing line.

Generally, the belt driven rotary vane type pumps operate at about 400 to 600 RPM and the direct drive pumps at about 1500 to 1700 RPM. Rotary vane type pumps can be considered as workhorse of vacuum. These pumps are commonly used in research and development laboratories and industry. These pumps find applications in many laboratory processes where pre-evacuation and backing pump action is required as in case of vapour diffusion pumps. These pumps are also employed in automotive industry and in industrial processes such as degassing, distillation, filtration and coating.

Figure 6.5 shows the rotary piston pump. The rotary piston moves in a circular path within the pumping chamber. As it passes top dead center, it creates a constantly increasing internal volume on the inlet end and a decreasing internal volume on the exhaust end.

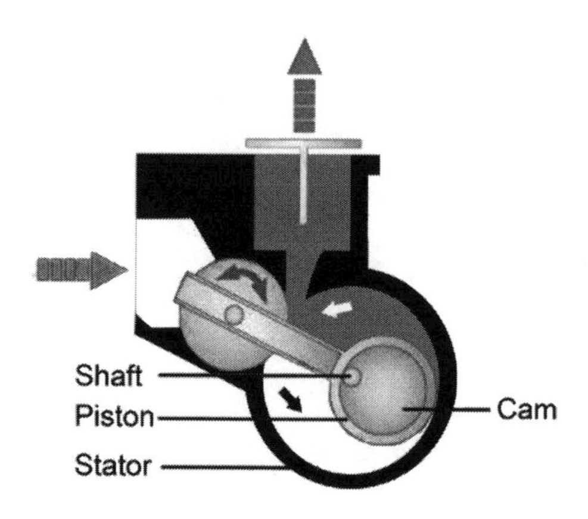

Fig. 6.5. Rotary piston pump *(Courtesy: Precision Plus).*

The moving piston drags the oil that forms a film into the clearance. High compression ratios are available in this pump. The pump offers high pumping speeds at low pressures. Pumps are available in either single-stage or two-stage design. The two stage pumps can achieve a better vacuum level due to the two piston chambers which are arranged in series with a connecting channel.

Oil-sealed rotary piston pumps are widely used for applications requiring vacuum over a wide pressure range. Due to the close tolerances between the stator and the rotor, the solid particulate matter entering the pump can cause damage to the sealing surfaces, thereby deteriorating the pump performance. This can be avoided by taking sufficient precautions including mounting of filter at the pump inlet.

On each cycle of the rotating eccentric piston and the sliding valve assembly, the gas is trapped into the stator, compressed, and further ejected to the atmosphere. Rotary piston type pumps can be single stage or compound. Rotational speed is about 600 to 800 RPM. The dimensional tolerances between the stator and the piston in pumps of this design are usually 0.08 to 0.1 mm. These pumps require higher viscosity oil due to the larger dimensional tolerances that are present. Large rotary piston pumps are often water-cooled to improve the pump life and performance.

Oil sealed rotary pumps can perform satisfactorily over long periods if serviced regularly that involves periodical change of the pump oil. Occurrence of vibrations and noise are generally associated with oil sealed rotary vane and piston pumps. These are minimized to a considerable extent in modern pumps.

6.1.3 Turbo-molecular Pump

In 1912, Gaede developed a simple molecular pump based on the principle that gases are caused to move in a preferred direction due to the interaction with high speed surfaces. This pump was similar in construction to a rotary vane mechanical pump, except that the rotor of this pump was without moving vanes and was concentric with the stator.

The turbo-molecular pump (TMP) is a dry compression type pump that uses a construction of rotor and stator blades in such a way that the gas molecules to be pumped receive momentum from the rapidly rotating blades. In the molecular flow region, the gas particles colliding with the moving blades are temporarily adsorbed on the surface and are then released. Thus the gas particles gain additional thermal energy from speed of the blades.

Becker[1] has described the theory and construction of axial flow turbo-molecular pump (TMP). Figure 6.6 illustrates the schematic of TMP by Becker. The pump consists of a horizontal cylindrical casing with series of annular rings of fixed vanes having oblique slots, uniformly spaced. Bearings at the end support a rotor shaft carrying a series of rings occupying the space between successive rings on the stator.

The vanes on the rotor are also slotted obliquely corresponding to the stator vanes as shown in Fig. 6.7. The vanes are a few mm thick and the rotor–stator clearance is about a few hundred to a few thousand μm. The pump employs about 10 rotor discs. A rotor disc and the adjacent stator disc constitute a pump stage that gives rise to a particular compression ratio. The rotor rotates at a speed of about 20 000 to 90 000 RPM. The rotor blades transfer momentum to the molecules in a preferred direction. The pump intake port is situated in the centre of the housing and the pumped gas is carried

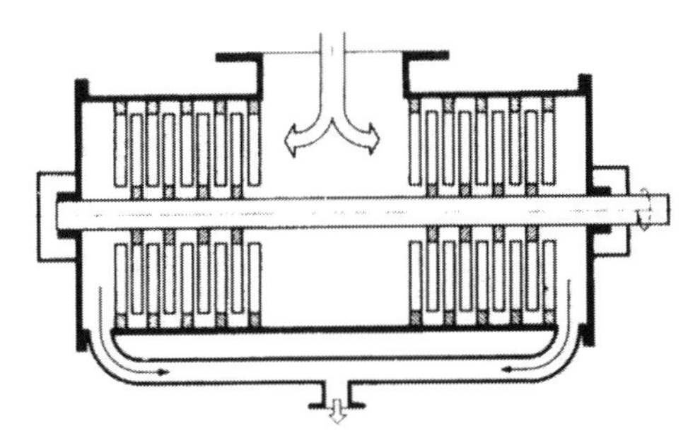

Fig. 6.6. Schematic of turbo-molecular pump by Becker *(Courtesy: CERN)*.

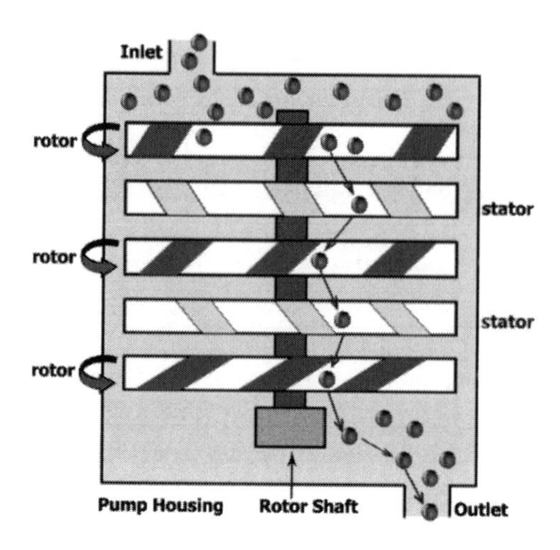

Fig. 6.7. Schematic of vanes structure of turbo-molecular pump.

to both ends to a backing pump. Oil-lubricated and cooled bearings are situated at the fore-vacuum side. Vanes are made of an aluminum alloy and the pump case of a stainless steel. Because the compression of each stage is ~10, each stage closer to the outlet is considerably smaller than the preceding inlet stages. The inlet stage is designed to achieve high speed with minimal compression. The stages close to the exhaust are designed to maximize compression of the gas. Compression ratios in excess of 10^{12} are achieved. The pumps are designed to operate in molecular flow conditions in the pressure range 10^{-8} to 1 Pa with foreline pressure from 10^{-1} to 10^4 Pa.

The later version of TMP is a multi-stage Gaede type molecular pump having a helical pump channel known as Holweck stage. Cylindrical sleeves that roate around helical channels in the stator are used as Holweck stages. Stators are designed to be located outside and inside the rotor to enable two Holweck stages to be integrated within the same pump. Thus the gas is displaced outside the rotor through the stator channel and inside the rotor through further stator channels till the gas is transported to the backing pump. Holweck stages are able to supply a high pumping speed, due to the presence of many channels in parallel and offer a high compression ratio.

Figure 6.8 shows the schematic of the Holweck stage of TMP.

TMP with speeds up to about 3 $m^3 \cdot s^{-1}$ are available. The pumping speed of the TMP for different gases varies with atomic weight. Figure 6.9 shows the variation of pumping speed of a typical TMP with pressure for different gases.

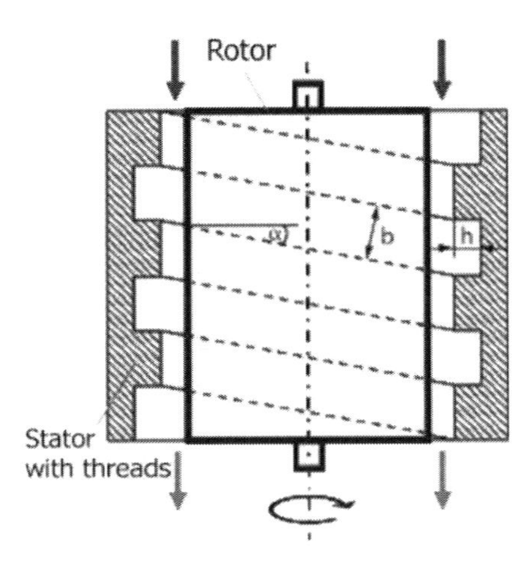

Fig. 6.8. Schematic of the Holweck stages turbo-molecular pump *(Courtesy: Pfeiffer Vacuum, "The Vacuum Technology Book, "Part 2: Know-how book", http://www.pfeiffer-vacuum.com).*

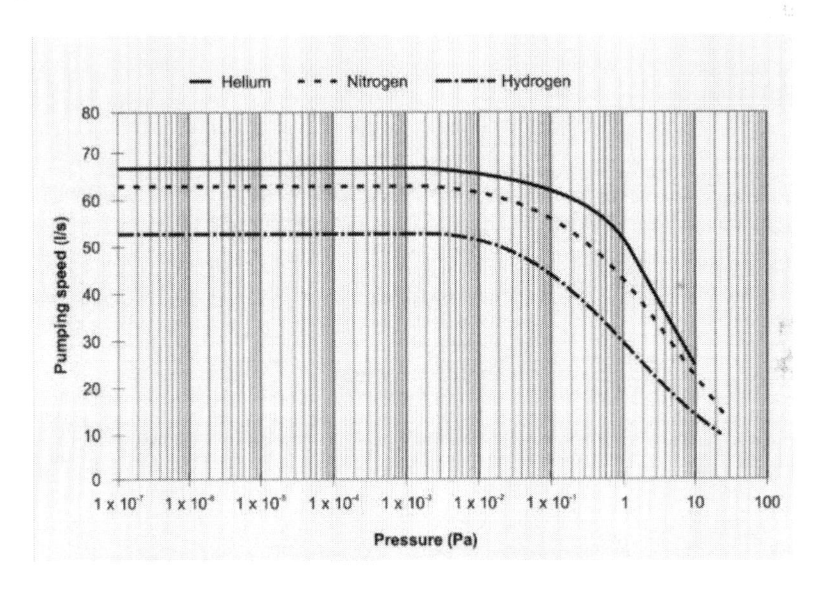

Fig. 6.9. Variation of pumping speed with pressure for a typical turbo-molecular pump.

The composition of the residual gases in TMP is influenced by that on the fore-vacuum side and the ultimate pressure is dependent on the outgassing from the large surface area of the rotor and stator blades. The pump is free from fluids, traps and baffles. The pump needs to be thoroughly outgassed for attainment of low pressures. The ultimate pressure of TMP is about 10^{-8} Pa. The residual gas pressure of a system pumped by TMP is mainly contributed by hydrogen. The major disadvantage of TMP is the low compression ratio for hydrogen. The fore-pressure of hydrogen necessary to maintain its partial pressure in high vacuum side to a value less than 10^{-9} Pa is about 10^{-6} Pa.

TMP cannot operate in the viscous flow regime. Hence, the rotary pump is employed for roughing from atmospheric pressure. The TMP is switched on at a pressure of about 10^{-2} Pa when it can impart momentum to molecules for their transportation towards the exhaust port of the pump.

In the molecular flow region, the maximum compression ratio K_0 at zero pumping speed rises exponentially with the square root of the molecular weight of the gas pumped with fixed circumferential speed u and geometry g given by

$$K_0 = \exp\left(M^{\frac{1}{2}} \times u \times g \right) \tag{6.1}$$

where

M: Molecular weight of the gas
u: Circumferential speed of the rotor
g: Geometry of the pump

This relation is valid only for an ideal TMP without losses.

TMPs find applications in particle accelerators, plasma fusion systems, evaporation chambers primarily due to their capability of offering organic contamination-free, clean vacuum. These pumps are employed in laboratory processes including sputtering, film deposition, ion implantation.

The rotor can be magnetically suspended to avoid lubricated bearings. With the low compression ratio for hydrogen of about 10^4, the conventional TMP can achieve an ultimate pressure about 10^{-3} Pa. Enosawa et al[2] developed a tandem TMP configuration with magnetic suspension having two TMPs on a common shaft. This pump offered a H_2 compression ratio of 5×10^8 with a maximum backing pressure of 50 Pa and achieved a pressure of 10^{-9} Pa. Cho et al[3] achieved pressure in the low 10^{-8} Pa range with a TMP using a magnetic suspension backed by a molecular drag pump and a dry diaphragm pump. After careful baking they achieved a lower ultimate pressure.

Lightweight devices help unmanned aerial vehicles to offer ease of remote handing and also allow vehicles to cover longer distances. Compact vacuum units that can be deployed would facilitate lowering the power consumption as these can be supported by smaller, lighter batteries to save more weight. Miniaturized, rugged, low-cost vacuum systems are being developed in response to requirements of NASA[4], other agencies and industry to replace the large, heavy and high-power consuming vacuum systems. A miniaturized vacuum system was developed based on a very small, rugged, turbo-molecular pump (TMP) which offers an order-of-magnitude reduction in mass, volume, and cost over current, commercially available, state-of-the-art vacuum pumps.

6.1.4 Roots Pump

Roots vacuum pump is a dry-running rotary displacement vacuum pump. It is also known as mechanical booster pump as it offers enhanced pumping speed for large gas loads in the pressure range between 1×10^{-1} Pa and 1 Pa where rotary vane and rotary piston pumps offer lower effective speeds. The upper pressure limit for this pump is about 2×10^3 Pa. In the Roots pump, two lobes of the figure-eight shape counter-rotate synchronously in a common housing as shown in Fig. 6.10.

The lobes are separated by a small narrow gap. The Roots vacuum pump can operate at high speeds (1500–3000 rpm) due to the absence of friction. The rotor shaft oil-lubricated bearings are positioned near the two ends of the drive shaft. The generated heat causes the lobes to expand more than the housing. The energy dissipation is limited by an overflow valve that prevents contact seizing, when the maximum

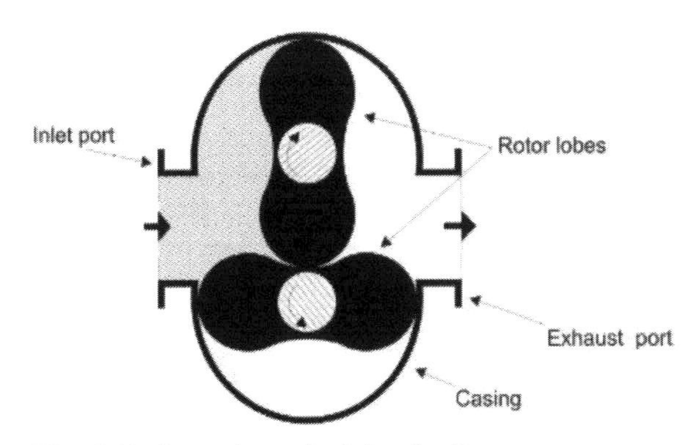

Fig. 6.10. Operating principle of a Roots pump.

possible pressure differential exceeds. It is coupled to the inlet wing and the exhaust wing of the pump. The gravity-loaded valve plate opens when the maximum pressure differential exceeds and facilitates a portion of the intake gas to flow back from the exhaust wing to the intake wing, based on the volume of gas. Alternatively, some pumps are gas cooled by a flow of gas from the exhaust port into the inlet chamber through a cooler. This enables the pump to compress and discharge against the atmospheric pressure.

The Roots pump is not capable of discharging directly against atmosphere and needs a backing rotary pump. The compression ratio is limited by

- the backflow of gas into the gaps between piston and housing
- the gas present on the surfaces of the piston on the exhaust wing is released upon rotating toward the intake wing.

The gap widths influence the compression ratio. These should be limited to certain minimum values considering the thermal expansion of the pistons and the housing. The small compression ratio of the pump varies with the molecular weight of the gas. Light gases possess better access to escape back from around the edges of the rotors. Thus it is essential to employ rotary pumps to back the Roots pumps. For a gap of 1.0–0.15 mm between the rotors, the typical speeds are about 1500–3000 rpm.

6.1.5 Diaphragm Pump

In this dry compressing type of pump, a diaphragm is placed between the pump head and the casing. A connecting rod that moves the diaphragm in an oscillatory motion causes the pumping action. The suction chamber comprises the space between the head and the diaphragm. Pressure-controlled shutter valves made of elastomer materials are used as inlet and outlet valves to control the flow of gas. The suction chamber is isolated from the drive by the diaphragm. The pumping chamber is exposed to the inlet as the chamber volume increases and is exposed to the exhaust line during compression. The diaphragm is coated with PTFE and exhaust valves use fluorinated elastomers. Thus, chemically active vapours gases can be pumped.

Figure 6.11 illustrates the principle of operation of diaphragm pump. V_1 and V_2 are the inlet and the outlet valves respectively. These valves operate resulting from the difference of pressure. As the diaphragm is lifted upwards the volume of the chamber increases

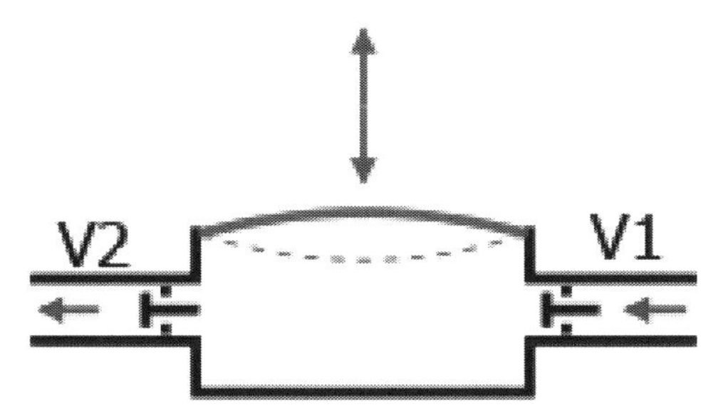

Fig. 6.11. Pumping action of diaphragm pump *(CC BY-SA 4.0 - Aflafla1 (Wikimedia Commons) – https://commons.wikimedia.org/wiki/File:Diaphram_pump.svg).*

resulting in lower pressure in the chamber, thereby resulting in opening of the valve V_1 while the valve V_2 remains closed. Lowering of the diaphragm increases the pressure in the chamber, causing the valve V_1 to close and the valve V_2 to open. Thus the gas entering the inlet port is transported through the chamber to be exhausted from the outlet port. The volume between the outlet valve and the suction chamber offers a limited compression ratio so that an ultimate pressure of approximately 7×10^3 Pa can be achieved. The ultimate pressure of 50 Pa is achievable with multiple pumping stages in series. The ultimate pressure attainable is limited by non-availability of sufficient force required to open the inlet valve. Pumping speeds of the order of 10 $m^3 \cdot h^{-1}$ are generally offered by diaphragm pumps. The diaphragm pumps can be used as backing pumps for TMPs. These pumps can be used for vacuum filtration/distillation/drying/degassing.

6.1.6 Screw, Scroll and Claw Pumps

The screw pump is a dry compression type pump based on the movement of screw. The housing of the pump encloses two synchronized screw rotors that counter rotate at about 3600 rpm. The pump offers progressive compression with tapers and variable pitch.

There is steady movement of the trapped gas from the intake to the exhaust wing of the pump resulting in pumping action as shown in Fig. 6.12. Pumping speeds of up to 9 m^3 per minute are offered by these pumps. The screw pumps offer an ultimate pressure in 10^{-1} Pa range. The pumps can operate at atmospheric pressure. The constructional materials are chosen to enable the screw pump to operate in the harsh environments of aggressive gases and particulates found in semiconductor etching and CVD processes. They are also used for roughing dry, high vacuum transfer or initial

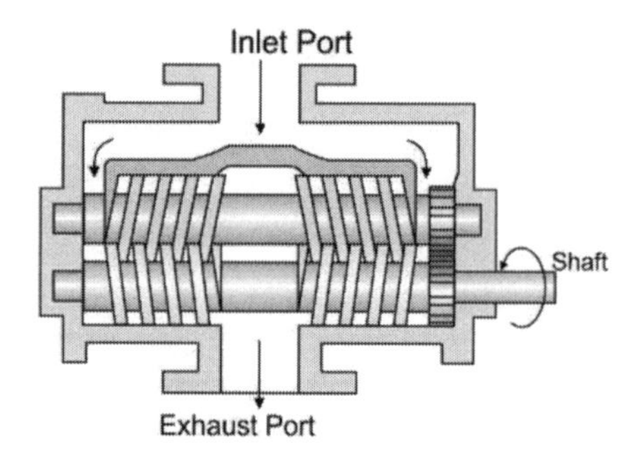

Fig. 6.12. Schematic diagram of screw pump.

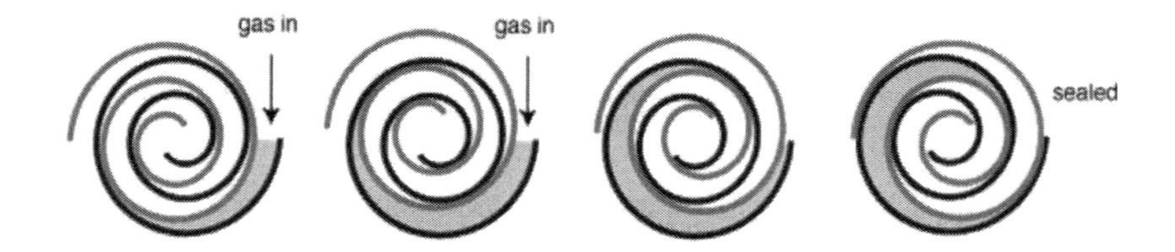

Fig. 6.13. Involutes of scroll pump *(Courtesy: Philip Hofmann).*

pumpdown for capture pumps. Mechanical dry-screw vacuum pumps are employed at several stages in the manufacture of Li-ion cells to deliver reliable, high-performance batteries with long lifetimes that can support electric vehicles. These pumps are used to vacuum-dry electrodes and complete battery cells to effectively remove moisture and solvents and also during electrolyte filling and degassing to stop micro-sized air bubbles from forming and to remove water vapour, in order to preserve efficient charging and discharging cycles.

The scroll pump uses two interleaved involutes of a circle of the same size resembling Archimedean spirals coupled with an offset of 180° thereby generating a number of crescent-shaped pockets. Using an eccentric drive, one spiral orbits around the fixed spiral without rotating. This action results in reduction in volume of the pockets thereby effecting progressive compression of the gas in one direction resulting in pumping action. Scroll pumps offer pumping speeds up to 52 m^3/hour. The ultimate pressure of the order of 8×10^{-1} Pa can be achieved with the scroll pump. Figure 6.13 illustrates the involutes of the scroll pump.

The scroll pumps find applications where contamination-free clean pumping is required. These include beam lines, electron microscopy. They are also used for backing TMPs.

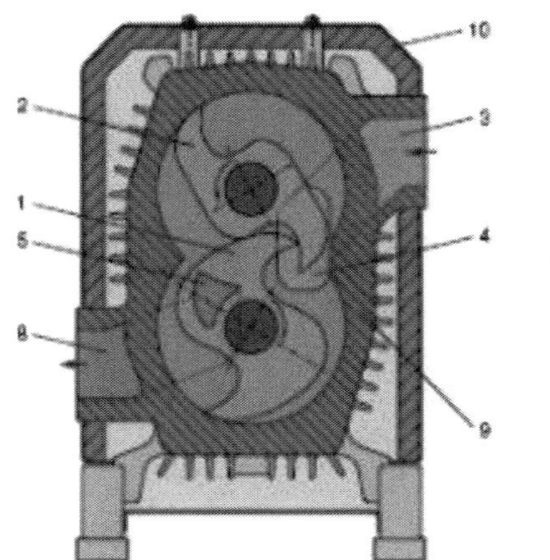

Fig. 6.14. Claw pump assembly *(Courtesy: Dynapumps)*.

Claw pumps are dry compressing type rotary piston vacuum pumps with two claw shaped rotors with their claws and the matching recesses counter-rotating. The clearance between the rotors and between the rotors and the wall of the casing is of the order of a few 0.01 mm. The intake and discharge slots are opened and closed periodically as the gas enters and gets trapped, compressed and later discharged into the next stage for pumping.

Figure 6.14 illustrates the assembly of the claw pump. The compression through the pump can be controlled by sizing the stages. The reversed claws of the pump allow a straight path. The claw pumps can offer speeds up to 550 m^3/h and the ultimate pressure of 2×10^3 Pa. The pump is capable of handling dust satisfactorily. The claw pumps are used in harsh industrial environments, particularly in semiconductor processing and where the water vapour content is high.

6.1.7 Diffusion Pump

The diffusion pump can be considered as the workhorse of high vacuum technology.

The diffusion pump consists of a hollow water-cooled body with a boiler at the bottom containing the pump fluid. The fluid is heated causing its evaporation. The high pressure (about 4×10^2 Pa in case of oil) fluid vapour is forced into the jet assembly mounted inside the body of the pump. The vapour is then streamed out of the jets at supersonic velocity. The diffusion of gas molecules through the low-density periphery of each jet into the dense cover of the

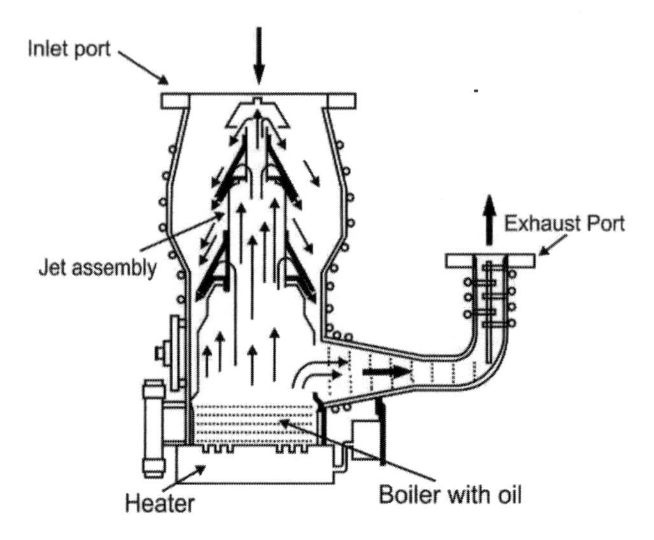

Fig. 6.15. Movement of vapours in oil diffusion pump.

freely expanding vapour stream results in pumping action. The annular jets are placed in vertical succession in such a way that the topmost jet offers a maximum conductance resulting from the largest annular cross-sectional area surrounding it. Each jet in the downward direction offers lower conductance as compared to the successive jet located above.

Working of an oil diffusion pump is shown in Fig. 6.15. The pump has three jets. The gas particles of the gas to be pumped get entrained into the vapour streams and gain velocity components in the direction of the exhaust port while the fluid vapours condense onto the cooled wall of the pump body and return to the boiler. The gas is successively compressed with each jet in downward direction depending upon the geometry of the pump body/jet assembly.

In some diffusion pump designs, the wall of the pump body is shaped to offer the required compression instead of using jets of different diameters.

The backing pump handles the pumped gas load to be discharged against the atmosphere. The diffusion pumps can offer pumping speeds up to 90 $m^3 \cdot s^{-1}$ and can operate in pressures in the range between 10^{-8} Pa and 10^{-10} Pa depending upon the pump fluid used, the traps employed and the processing of the system.

Figure 6.16 shows the variation of pumping speed with pressure of a typical oil diffusion pump. The pumping speeds start dropping at the high pressure end of the curve when the limiting fore-pressure

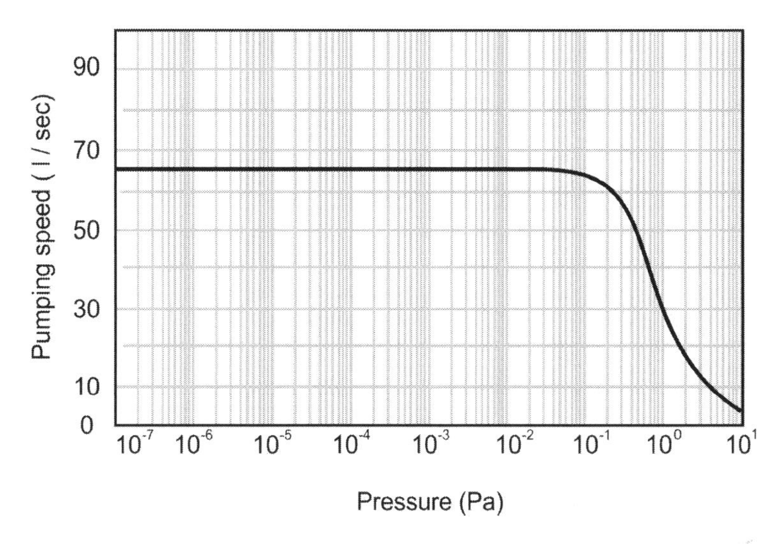

Fig. 6.16. Variation of pumping speed with pressure of a typical oil diffusion pump.

is approached as the vapour stream at the top jet is disturbed. The diffusion pump exhibits a constant speed throughout the low pressure range.

The actual pumping speed of a diffusion pump is less than its ideal pumping speed calculated from the conductance of the annular region at the top jet of the pump. The ratio of 'Actual Speed/Ideal Speed' is known as the Ho coefficient and normally lies between 0.4 and 0.5.

Apart from the outgassing from the walls, the diffusion pump encounters influx of gas as given below:

- Back-streaming of the vapours of the pump fluid into the system being pumped
- Back-migration due to evaporation of the condensed vapours and their entry into the system being pumped
- Back-diffusion of the pumped gas from the compressed state into the system being pumped.

The pump fluids for the diffusion pump include hydrocarbon oils, esters of phthalic, sebacic and phosphoric acids, silicone fluids. Among the fluids, silicone fluids are chemically and thermally stable and are widely use. Mercury is also used as a fluid, particularly in glass pumps. Its advantages include absence of decomposition, higher fore-pressure tolerance, ease of trapping, offering vacuum, free from organic contamination in the system. Its limitations include the necessity of refrigerating traps, the sensitivity to the surface condition of the pump and hazardous substance that can cause poisoning, if mishandled.

The cooled baffle valve which can isolate the diffusion pump from high vacuum chamber and also can intercept and condense vapours from diffusion pump is commonly used in diffusion pumped systems. The baffle valve is generally cooled by circulating chilled water around its exterior. The circular plate mounted internally serves as the baffle. This plate can be lowered to seal internally against the lower flange of the assembly for the isolation. Incorporation of 'cold caps' or cooled guard rings around the top jet has resulted in reduction of back-streaming and back migration of the fluid vapours as reported by Power and Crawley[5], Hablanian and Steinherz[6]. Water-cooled chevron baffles and liquid nitrogen cooled traps have proved much effective in minimizing back-streaming of fluid vapours. Alpert[7] used a copper foil trap at room temperature to reduce fluid vapours. Back-diffusion depends upon the compression ratio offered by the pump and falls rapidly for gases of higher molecular weights. Traps using the closed-loop refrigeration cycle are commercially

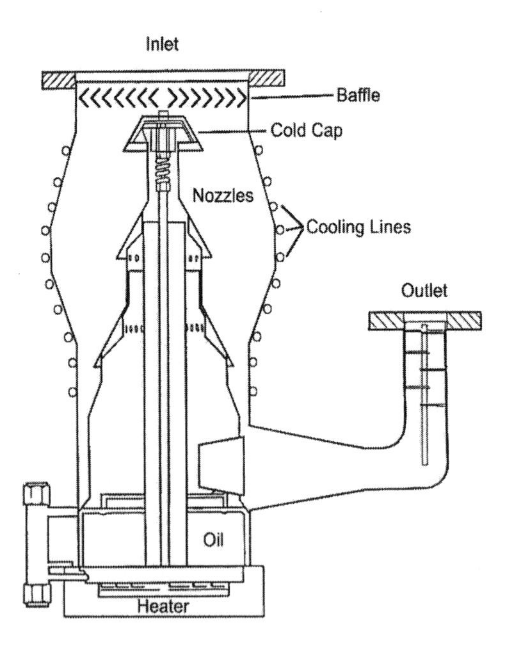

Fig. 6.17. Schematic diagram of diffusion pump with chevron baffle and cold trap *(Courtesy: Midwest Tungsten Service).*

available. These traps are convenient to use as one avoids frequent filling up of the consumable refrigerant.

The cold traps serve as cryogenic pumps. In the design of the traps, it is important to ensure that formation of a spot of intermediate temperature is avoided close to the vacuum side where condensation of water occurs, as this would cause slow evaporation of water in vacuum. Figure 6.17 shows the schematic diagram of the diffusion pump with a chevron baffle and a cold trap.

Some amount of hydrogen from the forevacuum side can diffuse back through the vapour jets thereby reducing the capability of hydrogen pumping. Careful trapping is essential to prevent the possible backstreaming of the diffusion pump fluid and the rotary pump oil to obtain clean, oil contamination-free vacuum. Santeler[8] has claimed that a pressure of 10^{-12} Pa can be achieved by using trapped oil diffusion pumps and a getter.

6.1.8 Vapour Booster Pump

The operation of the vapour booster pump is similar to that of the vapour diffusion pump but generates boiler pressures approximately ten times higher than is typical for the vapour diffusion pump. The high boiler pressure supplies a dense flux of vapour to ejector nozzles, specifically designed to increase the throughput of the pump. The gas to be pumped is trapped in the vapour stream in the first jet cap. The gas is further compressed and transferred to the next stage. The gas, then passes through the ejector nozzle into the condenser and ejected through the outlet to be handled by the backing pump. The ultimate pressure of these pumps is typically in the range 10^{-3} to 10^{-2} Pa. These pumps exhibit considerable pumping speed for permanent gases. Additionally, they are tolerant to high backing pressures. Vapour boosters are typically used in the 10^{-2} to 10 Pa range where primary pump combinations are often at their limit and ordinary diffusion pumps exhibit instability. Booster pumps can handle throughput up to 10 Pa · m^3 · s^{-1} at 10^3 Pa. Booster pumps are designed particularly to pumping of contaminated systems and processes with high gas loads of hydrogen, hence offer suitability for use in applications involving drying, distillation, degassing, metallurgical, thin films and chemical processes.

The steam ejector pump employs the Venturi effect as a high-pressure fluid is converted into a high-velocity jet which creates a low-pressure region. The mixed fluid comprising the gas to be pumped and the steam passes through the throat of the convergent-

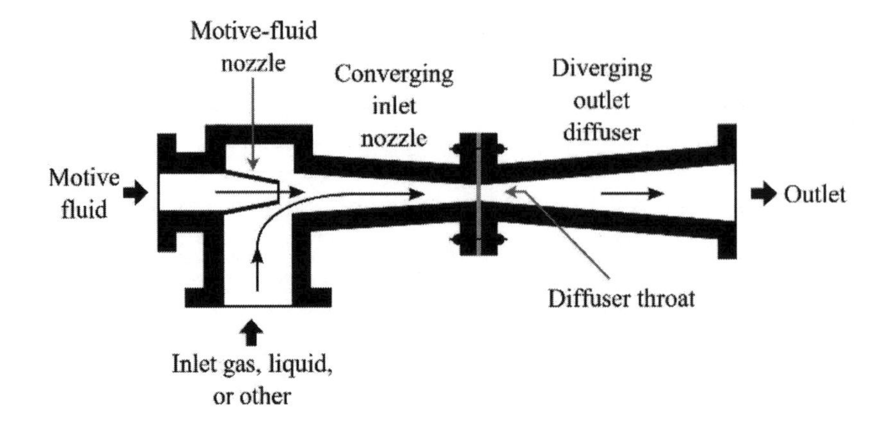

Fig. 6.18. Steam ejector pump *(CC BY 3.0 - Johannes Rössel (Wikimedia Commons) – https://commons.wikimedia.org/wiki/File:Ejector_or_Injector.svg).*

divergent nozzle. Upon its expansion, the velocity is reduced causing recompression of the mixed fluids.

Figure 6.18 illustrates the schematic of a single stage of a typical steam ejector pump. There are separate inlets for the gas to be pumped and for the steam. A number of pumping stages are generally utilized in the cascade. Two-stage pump combinations consisting of a steam ejector stage and a water jet (backing) stage are often used. The condensate from the steam is drained. The water jet stage of this pump is cooled with water to increase its efficiency. The steam ejector pumps are particularly suitable for pumping chemically active vapours. These pumps operate at a pressure of a few hundred Pa and are commonly used in the distillation equipment.

6.2 Pumps Using Conversion of Gas into Solid Phase

6.2.1 Getter Pumps

The gettering action involves one or combination of the phenomena including

- Chemisorption
- Bulk compound formation
- Solid solution
- Diffusion

Two types of getter pumps exist.

- Bulk or non-evaporable getters (NEG)
- Flash or evaporable getters

6.2.1.1 Bulk Getters

In bulk getters or non-evaporable getters (NEG), the chemisorbed gases and their compounds formed on the surface, diffuse into the bulk of the getter material at higher temperatures, thereby causing the pumping action. The diffused particles are trapped in the grain boundaries, interstitials and lattice defects. As the compounds formed at the surface diffuse into the bulk close to the surface, a fresh getter surface is exposed at the surface for further subsequent pumping. Alloys of Ti–Zr, Zr 84%–Al 16%, Zr–graphite are used in the form of ribbon, powder and pellets in these pumps[9]. The surface of a non-evaporable getter is covered with a thin, stable oxide film when exposed to air preventing the reaction of the atmospheric gases with the bulk of the alloy, thus maintaining the bulk of the alloy free from contamination.

The activation process is required to prepare the getter surface for pumping. Initial activation of getters at temperatures between 900–1000°C in vacuum is essential to remove oxides and other gases as the compounds formed near the surface of the getter diffuse faster in the bulk at these temperature. Additionally, dissolved H_2 that is saturated in the bulk of the material is released out of solid solution during the activation and is pumped by the external pump. Normal operating temperature of about 400°C is necessary for pumping action. Water can be pumped faster at 300°C. Getters are assessed by their sorption speed and the capacity for various active gases. Getter materials offer high surface reactivity and high diffusivity, to provide high sorption speed and a large capacity. At room temperature, limited bulk diffusion occurs for most of the gases that causes saturation of the surface of the getter, and prevents the pumping of the gas. The gettering speed is influenced by the extent of diffusion of the surface oxide film into the bulk of the getter. Degassing process employed in certain processes, itself is sufficient for partial activation. The pumping capacity for N_2, CO is about 2×10^{-3} $Pa \cdot m^3/gm$.

The pumping action of bulk getters has been discussed by Kindl[10] and della Porta et al[9]. Sheets or wires of zirconium and titanium can efficiently chemisorb gases only at temperatures higher than about 600–700°C. However, an alloy of Zr 84%–Al 16% prepared from a mixture of Zr_3Al_2 and Zr_5Al_3 is found to be active at about 300–400°C. The stress present in the boundary area in the two components of the alloy arising from their different lattice parameters, intermetallic compounds results into the activity. Hydrogen does not form a stable compound with the getter alloy but diffuses rapidly into the bulk of the active material, where it is stored as a solid solution. A given concentration of hydrogen inside the getter alloy corresponds to an equilibrium pressure of hydrogen, which is strongly dependent on temperature. Bulk getters can be used as heat controlled reversible pumps, which pump hydrogen at low temperature and release it at high temperature. NEGs are much effective in the extreme high vacuum (XHV) range because of their high pumping speed for hydrogen.

During the manufacturing process, the alloy is converted into a fine powder by the surface oxide. The powder is then compressed or coated onto surfaces of different containers or metallic strips. The assembly can then be mounted in the device to be gettered.

NEG pumps have found widespread acceptance in particle accelerators, allowing the achievement of the most demanding vacuum conditions in low energy heavy ions rings or in storage rings. NEG pumps have also been used in front ends, beam lines, insertion devices, electron sources and systems requiring high differential pumping such as electron cooling and gas-jet targets. The NEG cartridge is integrated with a small ion pump into one compact combination pump unit. The getter cartridge provides large pumping speed and capacity for gases, acting as the main pump for the removal of active gases (O_2, H_2O, H_2, CO, CO_2, and N_2). The ion pump has the ancillary task of removing chemically non active gases like argon and methane which are not removed by the NEG. Suetsugu et al[11] have discussed the design study of a distributed pumping system using NEG strips for the slender beam pipes of particle accelerators

The need to provide UHV conditions along the low-conductance accelerator vacuum chamber in order to guarantee the required beam lifetime prompted the use of NEG strip. Mounted in a side chamber along the main beam chamber, the NEG strip runs along the entire circumference of the storage ring providing UHV. NEGs that provide distributed pumping have been fruitfully utilized in parts of the system where high levels of desorption were induced by synchrotron radiation. Several facilities have adopted NEG to directly coat a large portion of the internal surface of the accelerators. The

NEG getter material is used as a sputter deposited layer onto the internal surfaces of an insertion device or a vacuum chamber of an accelerator. This improves the pressure in narrow gap devices and conductance limited chambers. Benvenuti[12] achieved pressure of 5×10^{-12} Pa in a 3 m long section of an accelerator ring using a 43.5 m long Zr–V–Fe NEG strip, a sputter-ion pump of speed $0.4\ m^3 \cdot s^{-1}$ and a titanium sublimation pump. A method of coating the interior of a stainless steel system with a thin film (~1 μm) of getter material consisting of Ti, Zr, V and their binary alloys deposited by sputtering was evolved by Benvenuti[13]. After an in situ bakeout at temperatures of 250–300°C, pressures of the order of 10^{-11} Pa have been produced by using this method. The use of NEG pumping devices in particle accelerators has also prompted their use in other large-size experimental physics machines such as nuclear fusion equipment (Tokamak). In this application NEG pumps, primarily in the form of getter wafer modules and panels, have been used mainly for their high efficiency to pump hydrogen and hydrogen isotopes.

6.2.1.2 Flash and Evaporable Getters

Chemisorption of gases on freshly deposited film of chemically active material in vacuum is utilized for pumping action in case of flash and evaporable getters.

Early forms of evaporable getters included pure barium encapsulated in small iron or nickel tubes, barium–thorium alloys and barium–strontium carbonate mixtures. These getters posed instability. Development of the $BaAl_4$ alloy at SAES getters in the early 1950s eliminated the instabilities. $BaAl_4$ is stable in air and found suitable for high volumes. The life of the vacuum tube was extended to thousands of hours using the $BaAl_4$ getter technology. Flash getters were commonly used in electron tubes and vacuum sealed-off devices. Flash getters are fired during the seal-off to pump the residual gases and the gases that might be released subsequently. Ba is often used as a flash getter.

Evaporable getters are used in many ultrahigh vacuum systems to serve as booster pumps. Titanium sublimation pumps (TSP) are used in combination with getter-ion pumps. TSP contain elements such as Ti, Zr, Mo, Ba, W, Al, Ta, V, Nb and Er. Selection of suitable evaporable getters is governed by factors such as ease of evaporation, bulk corrosion, and the formation of stable and low vapour pressure compounds, explosion hazards and cost. Ti has proved to be the most suitable evaporable getter. Ti (85%)–Mo alloy is commonly used for thermal stability, mechanical strength and lower power consumption. Ti wire wound over W is also used in small laboratory systems. Sublimation for gettering is achieved by Joule's heating or radiation or electron bombardment.

Herb and Davis[14] have described a titanium evaporation pump utilizing electron bombardment of molten titanium in the form of a wire. Resistance-heated sublimators and radiation-heated sublimators have been described by Warren et al[15], Harra and Snouse[16] respectively. In the simplest form of the getter pump, a loop of a Ti alloy wire is resistively heated to sublime Ti and deposit onto cooled walls of the envelope. For use in laboratory, a simple Ti evaporator can be made by winding a 0.25 mm dia. Ti wire interwoven with a 0.1 mm dia. W into a 0.25 mm dia. W support bent into a circle of suitable size. Getter pumps with speeds up to 60 $m^3 \cdot s^{-1}$ have been designed and constructed by Kuznetsov et al[17]. In the pressure region between 10^{-2} Pa and 10^{-1} Pa, most getter-ion pumps (to be discussed later) offer a fraction of their maximum speeds primarily because at these pressures the deposition rate of Ti films is not comparable to the residual gas impingement rate causing a reduction in the sticking probability. Since the resistive heating in the getter pumps can considerably enhance the sublimation and deposition rates, the sticking probability and the pumping speeds can be maintained close to the maximum value. Thus, the getter pumps can be utilized as booster pumps to handle the large gas loads.

Clausing[18] has measured the initial sticking probability for various gases on titanium films deposited under different conditions. He observed that the initial sticking probability for H_2 varied between 0.05 in high vacuum at 10°C for batch evaporation to 0.07 with

Fig. 6.19. Titanium sublimation pump *(Courtesy: Gamma Vacuum).*

continuous evaporation. These values were 0.08 and 0.20 respectively for N_2 and 0.38 and 0.86 respectively for CO. With increase in pressure, the impingement rate of the gas increases and therefore for a fixed evaporation/deposition rate of Ti, the sticking probability reduces. To avoid this, the evaporation rate can be increased up to a certain limit, as the pressure increases.

Gupta and Leck[19] have discussed evaluation of the titanium sublimation pump. Cryo-getter pumps in which the getter material is deposited on a cooled surface are effective for pumping H_2 which is normally a predominant gas in ultrahigh vacuum systems. McCracken[20] has described getter pumps utilizing resistively evaporated Ti deposited over a liquid nitrogen-cooled stainless steel surface. Prevot and Sledziewski[21] have described a titanium evaporation cryo-getter pump for nuclear fusion and space simulation experiments. They have also shown that TiO at low temperatures can pump argon.

The life of a getter pump is limited by the amount of titanium in the pump and its evaporation/sublimation rate. Cooling the deposited films of getter is essential for effective pumping. CH_4, C_2H_4 and H_2 are generated by getter pumps at hot titanium. It is essential to thoroughly degas the titanium filament before using for its pumping action. Figure 6.19 shows a typical TSP.

Pumping Speed of Getters

We have seen that the rate dn_a/dt of capture of residual gases at pressure P and temperature T is given by equation (3.7)

$$\frac{dn_a}{dt} = s_t P(2\pi mkT)^{-\frac{1}{2}}$$

Here s_t is the sticking probability; P is pressure; m is the mass of the gas; k is the Boltzmann constant; T is the temperature of the gas in Kelvin.

Using equation (1.10), we have

$$\frac{dn_a}{dt} = s_t \times 2.63 \times 10^{24} P(MT)^{-\frac{1}{2}} \text{ m}^{-2} \cdot \text{s}^{-1}, \text{ if } P \text{ is in Pa}$$

As the pumping speed S is given by

$$S = \frac{1}{n}\frac{dn_a}{dt} \tag{6.2}$$

$$S = \frac{\left(s_t \times 2.63 \times 10^{24} P(MT)^{-\frac{1}{2}}\right)}{n} \mathrm{m}^3 \cdot \mathrm{s}^{-1}, \text{ if } P \text{ is in Pa}$$

$$= \left(s_t \times 2.63 \times 10^{24} P(MT)^{-\frac{1}{2}}\right) \frac{kT}{P}$$

$$= s_t \times 2.63 \times 10^{24} \times kT \times (MT)^{-\frac{1}{2}}$$

Substituting value of $k = 1.37 \times 10^{-23}$ Pa · m^3 · K^{-1},

$$S = s_t \times 2.63 \times 10^{24} \times 1.37 \times 10^{-23} \left(\frac{T}{M}\right)^{\frac{1}{2}}$$

$$= s_t \times 36 \times \left(\frac{T}{M}\right)^{\frac{1}{2}} \ \mathrm{m} \cdot \mathrm{s}^{-1} \tag{6.3}$$

where n is the gas density, M is the molecular weight.

Thus, for N_2, at room temperature (293 K), assuming $s_t = 1$, the pumping speed per meter square for the freshly deposited titanium film is 116.45 m · s^{-1}. Practically, as the value of s is about 0.5, the maximum pumping speeds of the order of about 5.8 liter · s^{-1}· cm^{-2} are achieved for most of the chemically active gases such as H_2, H_2O, CO, CO_2, O_2, and N_2. The getter pumps cannot pump rare gases and hydrocarbons. These gases can be removed by using the getter pumps in combination with other pumps at room temperature or liquid nitrogen temperature and pressures of the order of 10^{-11} Pa can be attained. Freshly deposited titanium films release methane unless high purity titanium is used or the film is baked at 100°C for a few hours.

6.2.2 Getter-Ion Pump

In getter ion pumps, ion pumping by entrapment of ions is involved in addition to getter pumping as the inert gas cannot be pumped by getter pumps alone. These pumps generally incorporate configuration of ion gauges for ionization of gases either by using hot cathode or by using cold cathodes as in case of sputter-ion pumps to be discussed later. The first ion pump using titanium evaporation was developed in Herb's laboratory[22] at University of Wisconsin in 1961. The pump used cartridges of titanium mounted around the circumference of a rotating wheel powered by magnet. The cartridge was heated by electron bombardment. A grid was incorporated to ionize gases caused by electrons sourced from another filament. The ions are trapped onto the surface of the pump wall where titanium is deposited. The pump offered a pumping speed of 1 $m^3 \cdot s^{-1}$ for air and of 0.104 $m^3 \cdot s^{-1}$ for argon.

Figure 6.20 illustrates one of the initial getter-ion pumps developed by Herb[23].

Redhead et al[24] developed a pump using the inverted magnetron. The pump has anode in the form of a helical coil of titanium wire wound together with tungsten wire mounted on a sapphire rod along the axis of a cathode in form of a cylindrical box. The axial magnetic field of about 2000 Gauss is applied and the anode voltage is 6000 V. Titanium is deposited on the inner wall of the cathode where the ions also are entrapped. The getter-ion pumps based on conventional triode and tetrode electron tubes configurations were built by Huber and Warnecke[25].

Orbitron pumps were developed in Prof. Herb's laboratory[26] at University of Wisconsin. The pump is based on the configuration of the orbitron gauge discussed earlier. The performance of a 5-cm diameter glass orbitron pump shown in Fig. 6.21 was studied by Naik and Herb[27].

Electrons injected by the biased filament in the radial electrostatic field between two concentric cylinders hit the titanium cylinders.

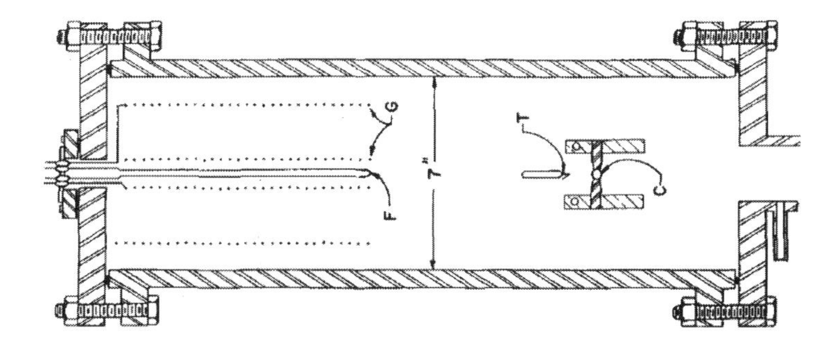

Fig. 6.20. Getter-ion pump *(Courtesy: Servicos Source: Electrostatic Accelerator Development at Wisconsin, R. G. Herb, Revista Sociedade Brasileira de Física, Vol. 2, N." 1, 1972).*

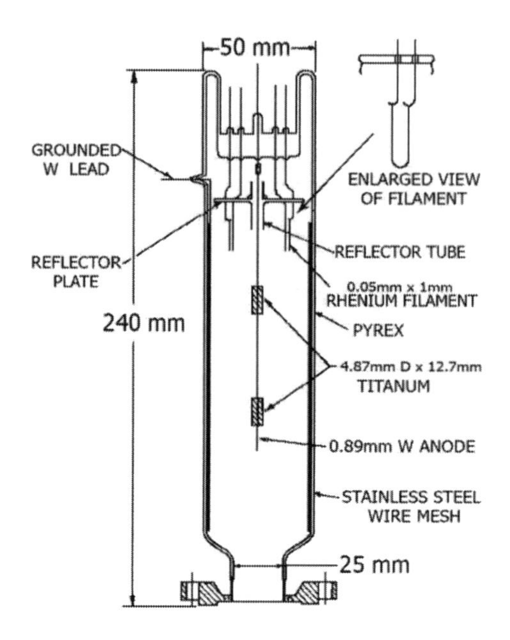

Fig. 6.21. Glass orbitron pump *(Reproduced with permission from P. K. Naik and R. G. Herb – J. Vac. Sci. Technol. 5, 42 Chapter 6, (1968), American Vacuum Society).*

Those with greater angular momentum continue to spiral around the central anode. The titanium is heated to the sublimation temperature by electron bombardment and chemically active gases are pumped by fresh deposit of titanium on the wall of the pump. Inert gases, ionized by orbiting electrons, are driven to the walls and buried there by fresh titanium. The pumping speed for chemically active gases is limited by the throat impedance if the sublimation/deposition rate of titanium is sufficiently high and the sticking probability remains close to unity.

A 10 cm in diameter orbitron pump offered pumping speeds of 0.9 $m^3 \cdot s^{-1}$ for hydrogen, 0.5 $m^3 \cdot s^{-1}$ for nitrogen, 0.3 $m^3 \cdot s^{-1}$ for air, and 7×10^{-3} $m^3 \cdot s^{-1}$ for argon.

A modified orbitron pump developed by Naik and Verma[28, 29] employs two filaments. One of the filaments injects electrons to sublime titanium while the other filament is used to inject electrons of large average path to ionize the gas effectively. A grid mounted coaxially is biased suitably to impart additional energy to the positive ions that are efficiently trapped onto the continuously replenished titanium surface. Figure 6.22 shows the modified orbitron pump. The pump offers 47% to 66% increase in the pumping speed of helium over the conventional orbitron pump.

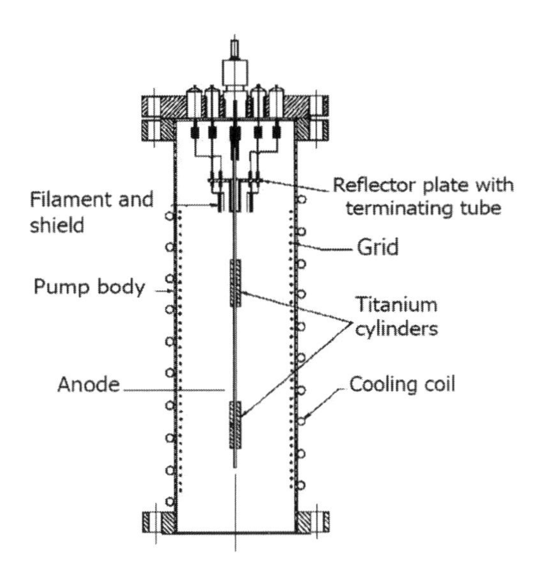

Fig. 6.22. Modified orbitron pump *(Reproduced with permission from P.K. Naik and S.L. Verma- J. Vac. Sci. Technol., A 14, 734 (1977), American Vacuum Society).*

6.2.3 Sputter-Ion Pump

The phenomena that play important role in the mechanism of pumping in the sputter-ion pump (SIP) include

- Penning discharge ionization
- Sputtering
- Deposition of metal films
- Chemisorption/Gettering
- Ion burial/pumping
- Entrapment of energetic neutral gas particles in metals
- Back-scattering of ions

6.2.3.1 Normal Diode Pump

A diode SIP[30] consists of a multiple anode cells structure between two cathode plates with the magnetic field in a direction shown in Fig. 6.23. A number of Penning cells are put together in its structure. The hollow anode is cylindrical and made of stainless steel. The cathode plates are made of titanium. The anode is held at positive voltage, few kilovolts, with respect to the grounded cathodes and a magnetic field of 800–2000 Gauss is utilized. The pump configuration presents a crossed electric and magnetic field

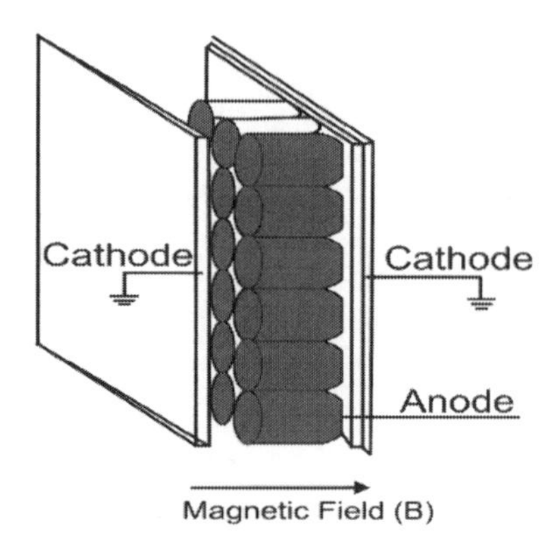

Fig. 6.23. Normal diode sputter-ion pump.

arrangement resulting from formation of a negative space charge along the axis of each Penning cell. The average path length of electrons is therefore increased to cause significant ionization of the residual gas. The positive ions are accelerated by the electric field to strike the titanium plates positioned on both sides of the anode. Upon striking the cathode plates, the ions cause

- Penetration and entrapment into the cathodes
- Sputtering of the cathode surface
- Back-scattering after neutralizing at the cathodes

The ions striking the cathodes penetrate to the extent of a few atomic layers and are trapped there causing pumping action for all kinds of ions. Such a pumping action is particularly important for ions of noble gases which cannot be pumped otherwise effectively. However, this pumping effect is not permanent since erosion of the cathode material from sputtering releases some of the ions trapped earlier. Heavy ions are trapped close to the surface of cathodes and are released when their concentration in the first few layers increases. Nevertheless, an effective ion burial does occur at certain locations on the cathode surface where there is a net build-up of the sputtered deposit from other regions of the cathode. This explains why the gases such as argon, methane are removed in a diode pump.

Positive ions striking the titanium cathode surface, sputter titanium atoms, forming deposit of titanium film on the surrounding surfaces, particularly on the large surface of the anode. This results in the gettering action in which the active gases such as N_2, O_2, H_2, CO, CO_2 and water vapour combine chemically with the freshly deposited titanium to form stable chemical compounds, thus giving rise to pumping. The pumping speeds, particularly for chemically active gases depend on the rate of deposition of sputtered titanium atoms which is proportional to the pressure inside the pump and the rate of sputtering which in turn is influenced by the mass, direction and energy of the bombarding ion. Thus, heavier gases cause a sufficiently high rate of sputtering while lighter gases such as hydrogen and helium offer a negligibly small rate of sputtering of titanium. However, hydrogen can form solid solution with titanium. Hydrogen ions can also diffuse deep into the bulk of the cathode and remain trapped.

The pumping speed of a single Penning cell of SIP depends on several parameters including its diameter, length in addition to electrical and magnetic fields. The pumping speed S that is available for the unit cell is given by equation 6.2

$$S = \frac{1}{n}\frac{dN}{dt}$$

where N is the number of gas particles

$$\frac{dN}{dt}\text{:Capture rate}$$

The term $\frac{dN}{dt}$ can be considered proportional to $\frac{I_p}{q}$ where I_p is the discharge current in A, q is the charge on the ion in Coulombs 1.6×10^{-19} Coulombs.

Thus

$$\frac{dN}{dt} = \beta\frac{I_p}{q}$$

where β is a constant, and

$$S = \left(\frac{1}{n}\right)\beta\frac{I_p}{q} = \left(\frac{kT}{P}\right)\beta\frac{I_p}{q} = \left(\frac{kT}{q}\right)\left(\beta\frac{I_p}{P}\right)$$

where k is the Boltzmann's constant, 1.37×10^{-23} Pa · m^3 · K^{-1}, P is the pressure in Pa, T is the temperature in K.

Thus

$$S = \left[\frac{1.37\times10^{-23}}{1.6\times10^{-19}}\right]\beta\frac{I_p T}{P}\ \mathrm{m^3 \cdot s^{-1}} =$$

$$= 8.5 \times 10^{-4} \, \beta \frac{I_p T}{P} \, \text{m}^3 \cdot \text{s}^{-1} \qquad (6.4)$$

where β corresponds to the fraction of ions pumped, I_p/P corresponds to the discharge intensity in current in Amperes per unit pressure. The discharge intensity is fairly constant in the pressure range 10^{-6}–10^{-2} Pa. Thus, a rough indication of pressure can be obtained from the discharge current I_p.

Inert gas pumping mechanism, argon instability

Helium is pumped through ion burial in the cathodes followed by limited diffusion in the bulk and is not released back. About 50% of the neutral helium atoms backscattered from cathodes are unaffected by the electric and magnetic fields and possess energies in excess of about 70% of the incident ion energy. These atoms are trapped onto the anode. The ratio of pumping speed S_{He} of helium to that of nitrogen S_{N_2} (S_{He}/S_{N_2}) is about 0.1 while the ratio (S_{Ar}/S_{N_2}) for argon is about 0.01 to 0.02.

Argon instability is observed when the diode pump is pumping on an 'air leak' or argon rich mixture of gases. The high rate of ion incidence in the centre of the standard Penning cell significantly erodes the cathode material and previously buried argon atoms are released. The pump suddenly releases part of the buried atoms in repeated bursts of pressure. Further developments in the SIP were aimed at elimination of argon instability and improvement of inert gas pumping speeds.

6.2.3.2 Stabilized Diode Pumps

6.2.3.2.1 Slotted Cathode Pump

Jepsen et al[31] developed a stabilized diode pump with slotted cathodes. The cathode surfaces, in this case, are grooved in such a manner that the slot width is less than the slot depth; the slat width is less than the slot width as illustrated in Fig. 6.24. Ions incident normally in the slot are trapped without causing much of sputtering.

Ions incident on the slats cause sputtering of the corners and the material is deposited in the slots to cover and bury the trapped ions. Thus the ion burial occurs at the regions where there is a net build-up of the sputtered material. It is experimentally observed that the grooving helps only along the outer regions of the cathode cell.

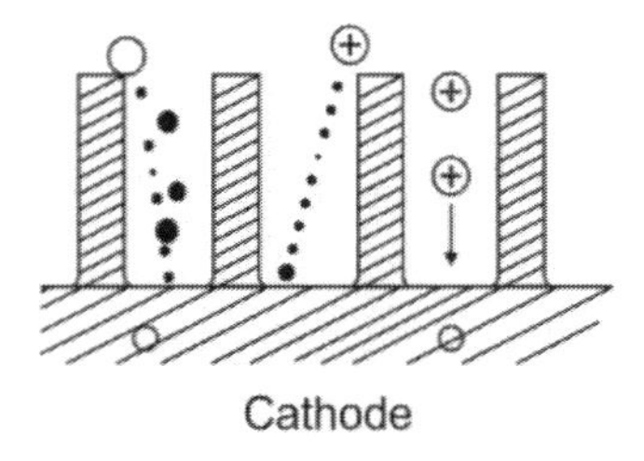

Fig. 6.24. Slotted cathode of sputter-ion pump.

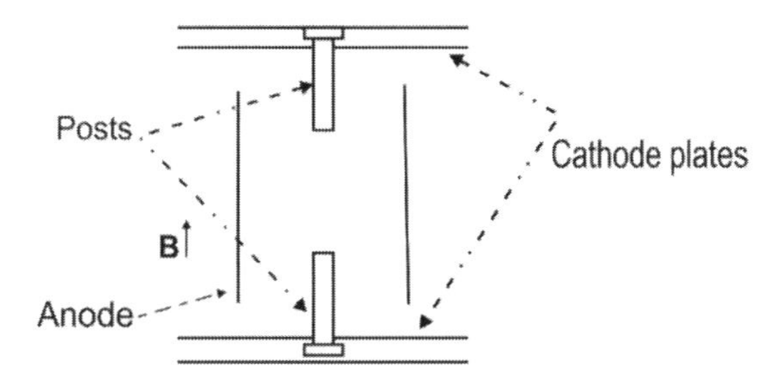

Fig. 6.25. Post cathode sputter-ion pump.

This change resulted into the reduction of argon instability and an increase in the pumping speed of argon. The ratio S_{Ar}/S_{N_2} is 0.05–0.1 for this pump.

6.2.3.2.2 Post Cathode Pump

Lamont[32] has described a post cathode diode pump which combines magnetron configuration with the normal diode. In this pump, the cathodes project axially, but not fully, into the anode cell as shown in Fig. 6.25. Sputtered material from the tops and the sides of the posts deposits on the anode assembly and on the cathode planes at higher rates than in the normal diode due to enhanced sputtering at glancing angles. The region for the net build-up of materials is thus enlarged and the pumping by ion burial is enhanced.

In this pump, back-scattering in form of energetic neutrals takes place dominating on the tops and the sides of the posts and to a lesser extent on the planar cathodes. These energetic neutrals can be pumped on the anode assembly as well as regions of net build-up on the planar cathodes. By utilizing tantalum as the post material, it was possible to increase the speed of argon so that (S_{Ar}/S_{N_2}) was about 0.25. A few percent rise in the speeds for chemically active gases has also been claimed. Compared to the normal diode, the discharge initiates easily at low pressures in this pump. The pump is useful as a small appendage pump.

6.2.3.2.3 Differential Ion Pump

The differential ion pump[33] (D-I Pump) uses one cathode of tantalum while the other cathode is made of titanium. The probability of back-scattering is a function of the mass ratio of the ion species and the cathode material and also on the angle of incidence of the ions striking the cathode surfaces. The incident ion is neutralized at the surface of titanium cathode. The penetrating probability of the backscattered neutrals is increased when the cathode material is changed over from titanium to tantalum cathode. Tantalum, also a chemically active material has a much higher atomic mass as compared to titanium (Ta:181 amu, Ti:48 amu). There are more elastic collisions of ions that strike the tantalum surface. The backscattered neutrals retain most of their initial kinetic energy and can be trapped deep in the anode or in the pump wall. Such a pump is commercially referred to as a 'noble diode' or 'differential ion' (D-I) pump and offers higher pumping speeds for noble gases. The ratio S_{Ar} / S_{N_2} is about 0.20 for this pump. However, this increase in the ratio is obtained at the cost of reduction of the pumping speed for chemically active gases.

6.2.3.3 Triode Pump

In the triode pump[34] the cathode plates are replaced with a grid made of titanium and are maintained at negative voltage with respect to the anode and the auxiliary electrode is maintained at ground potential as shown in Fig. 6.26.

In the pump, the ions bombard the cathode grid at glancing angle, giving rise to a high sputter yield of titanium. The energetic neutrals

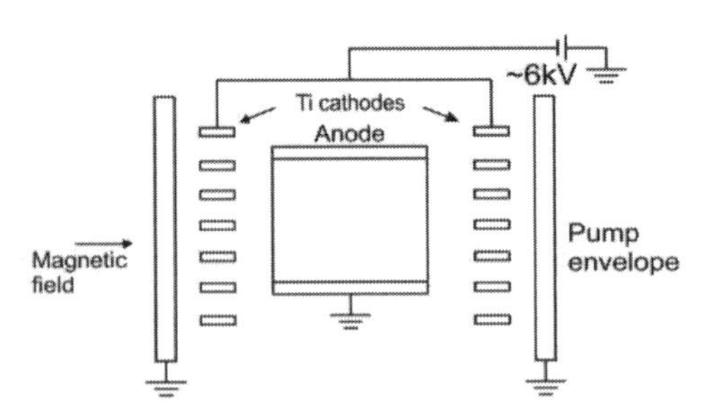

Fig. 6.26. Triode sputter-ion pump.

generated at the cathode grids strike the auxiliary electrode or are reflected back to be buried in the anode. The ions cannot reach the pump body or anode and the trapped inert gas atoms can remain buried on these surfaces. Thus stable pumping of inert gases is available. The pumping speed for inert gases is up to 20–25% of the air pumping speed. The reduction in the total pumping speed for chemically active gases is about 20% as compared to the standard diode.

6.2.3.4 Integrated Linear Pumps

Distributed sputter-ion pumps[35] as shown in Fig. 6.27 are used in particle accelerators, particularly in storage rings. The pump components are integrated into vacuum chamber of the bending magnets that provide the magnetic field for the pumps. The cylindrical anode cells are formed by stacked stainless-steel plates with coaxial holes. In electron storage rings, the beam induced gas desorption is significant and its intensity is governed by the beam energy that is decided by the field of the bending magnets which also governs pumping speed of the distributed sputter-ion pumps. Use of distributed sputter-ion pumps is best suited for the uniform desorption rate across the length. SIPs with combination of non-evaporable getters (NEG) pumping has been used in particle accelerators.

6.2.3.5 Pressure Range and Operational Aspects

The pumping speed of sputter ion pumps varies with pressure. The operating pressure is in the range of less than 10^{-2} Pa since at higher pressures, there is a change in the discharge mode that prevents the sputtering process. The maximum pumping speed is attained at about

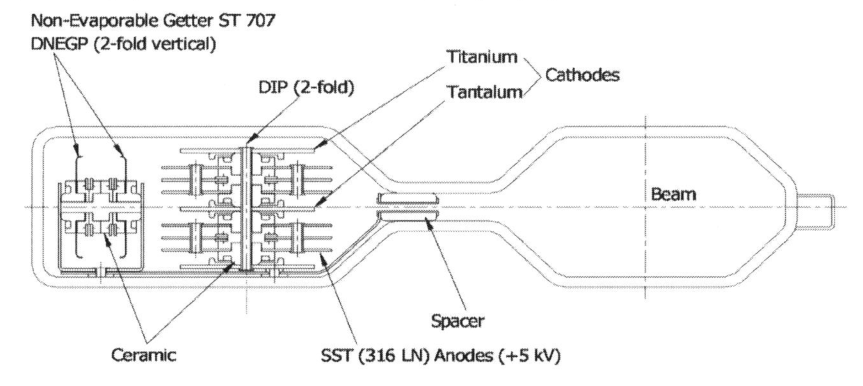

Fig. 6.27: Distributed sputter-ion pump in storage ring of particle accelerator *(Courtesy: L. Schulz, Paul Scherrer Institut).*

10^{-4} Pa. The pumping speed falls at lower pressures as the discharge intensity, current per unit pressure (I/P) decreases. The rate of fall in the pumping speed of an SIP at lower pressures depends on magnetic field, anode voltage, and dimensions of the pumping cell. Pumping speed is proportional to discharge intensity which increases with anode diameter. Increasing the magnetic field raises discharge intensity at low pressures.

The ultimate lowest pressure range of 10^{-9} and 10^{-8} Pa can be attained only after a bakeout of the pump. The lifetime of the pumps is decided by the operating pressure. They have longer life when operated at lower pressures. It is therefore advisable to start the sputter-ion pumps at pressures less than 10^{-2} Pa. The lower pressure limit of sputter-ion pumps is in the range of 10^{-9} to 10^{-8} Pa. The discharge current measured in a sputter-ion pump is indicative of the pressure. Re-emission of previously pumped rare gases occurs in SIPs.

6.2.4 Cryogenic Pumps

Cryogenic pumping/Cryo-pumping includes pumping of gases by

- Cryo-condensation
- Cryo-trapping
- Cryo-sorption

During cryo-pumping, the gas particles undergo direct phase transition from the gaseous to the solid phase without passing through the liquid phase. Pumping of residual gases by condensation on cooled surfaces is generally called cryo-condensation pumping. Cryo-trapping involves concurrent pumping of two or more gases by entrainment of gas particles which cannot be condensed at the prevailing temperatures and pressure conditions. A condensing gas is used for entrainment to form a mixed condensate. The smaller size molecules are trapped in the cryo-deposit of the majority species and are subsequently buried. Removal of gases by physical adsorption on high surface area and high porosity surfaces at cryogenic temperatures of certain substances is involved in cryo-sorption pumping. Day[36] has published an excellent review on cryo-pumping.

6.2.4.1 Cryo-condensation Pumping

For cryo-condensation, the surfaces are required to be cooled to a temperature so as to keep the corresponding saturation pressure equal to or lower than the desired vacuum pressure in the chamber. The achievable pressure is determined by the saturation pressure at the temperature chosen for the cold surfaces. The limiting pressure for the cryo-condensation pump is the vapour pressure of the adsorbed gas at the temperature of the surface, and its capacity is limited by the thickness of the condensed layer. With the typical condensation coefficients of about 0.5, maximum pumping speeds of about 5 liters · s^{-1} · cm^{-2} can be achieved. However, practically, such speeds are not possible because the cryo-surface is required to be shielded from room temperature radiation by suitable baffles at intermediate temperatures. Water and hydrocarbons can be condensed at the temperature of about 100 K while, the air components can be condensed at 20 K. Hydrogen isotopes and neon can be condensed at temperature of 4 K.

On a cryo-surface, capture rate = condensation rate – evaporation rate; capture rate = capture coefficient (c_c) × gas impingement rate, and condensation rate = condensation coefficient (c_0) × gas impingement rate; evaporation rate = evaporation coefficient (c_e) × maximum evaporation rate

Using equation (1.9), we have
The capture rate

$$\frac{dn_c}{dt} = c_c P_g \left(2\pi mkT\right)^{-\frac{1}{2}} \tag{6.5}$$

where P_g is the pressure of residual gas at temperature T; m, k and T are as defined earlier.

And

$$\text{Condensation rate} = c_o P_g \left(2\pi mkT\right)^{-\frac{1}{2}} \tag{6.6}$$

$$\text{Evaporation rate} = c_e P_s \left(2\pi mkT_s\right)^{-\frac{1}{2}} \tag{6.7}$$

where P_s is the vapour pressure of the condensate at temperature T_s. Thus

$$c_c P_g \ (2\pi mkT)^{-\frac{1}{2}} = c_0 P_g \ (2\pi mkT)^{-\frac{1}{2}} - c_e P_s \ (2\pi mkT_s)^{-\frac{1}{2}} \tag{6.8}$$

And

$$c_c = c_o - c_e \left(\frac{P_s}{P_g}\right)\left(\frac{T}{T_s}\right)^{\frac{1}{2}} \tag{6.9}$$

For $P_g \gg P_S$, $c_c = c_0$

Therefore,

The capture rate

$$= c_c P_g \ (2\pi mkT)^{-\frac{1}{2}} = c_0 P_g \ (2\pi mkT)^{-\frac{1}{2}} \tag{6.10}$$

Pumping speed S per unit area is given by

$$S = \frac{1}{n}\frac{dn_c}{dt} \tag{6.11}$$

$$= \frac{1}{n}\left[c_0 P_g \ (2\pi mkT)^{-\frac{1}{2}}\right] \tag{6.12}$$

where n is the gas density

$$S = c_0 \times 36 \times \left(\frac{T}{M}\right)^{\frac{1}{2}} \tag{6.13}$$

With $c_0 = 1$, for water vapour at room temperature on a liquid nitrogen cryo-panel, the maximum pumping speed S_w per meter square for water vapour is given by

$$S_W = 143 \ \mathrm{m^3 \cdot s^{-1}}$$

The cryo-pumps usually utilize pumping arrays. A pumping surface at 4–29 K generally uses shield or shrouds at 77 K or 100 K. This helps to shield the pumping surface from direct radiation and also the gas is pre-cooled before pumping to reduce the load on the actual pumping surface though this results into lowering of the pumping speed.

The active pumping surfaces, the thermal shields and the baffles are made of copper for better thermal conductivity. Figure 6.28 shows a typical arrangement of arrays for cryo-pumping.

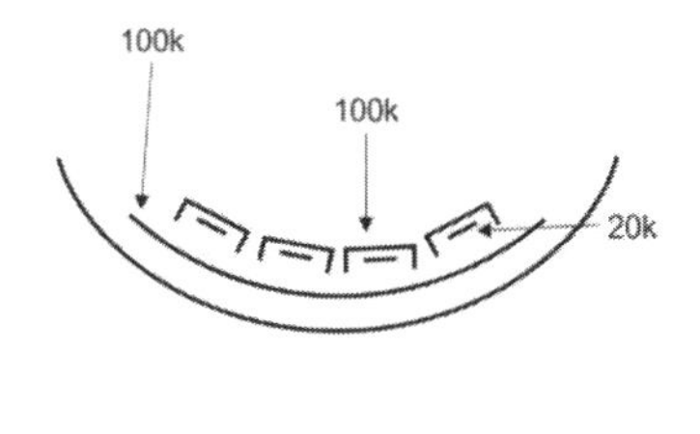

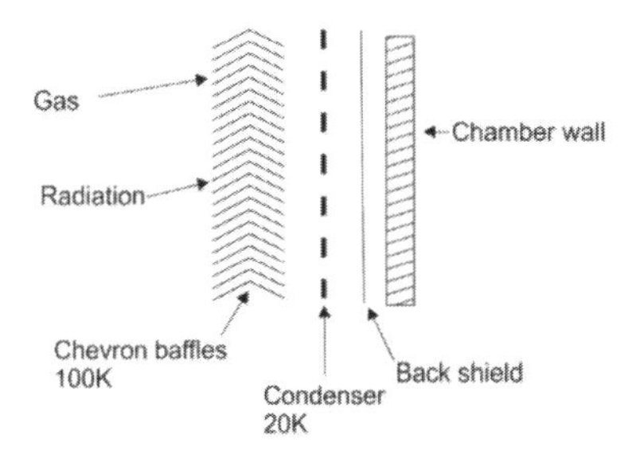

Fig. 6.28. Typical cryo-pumping arrays.

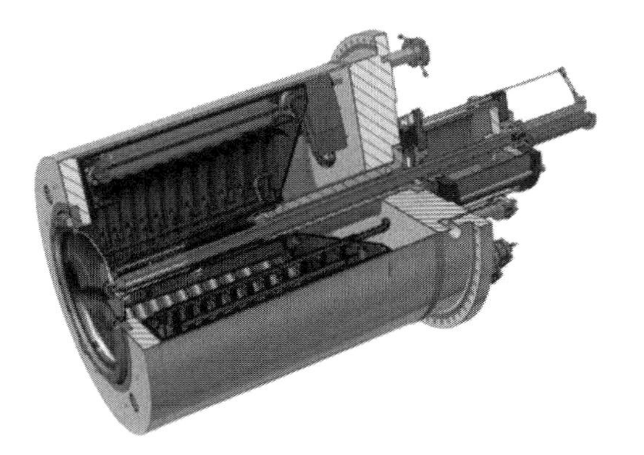

Fig. 6.29. Cryo-pump at ITER *(Courtesy: ITER).*

Figure 6.29 shows the cryo-pump at the International Thermonuclear Experimental Reactor (ITER).

At high pressures in the transition flow region, there exists a temperature and pressure gradient between the cold pumping surface and the walls of the containing vessel. This situation forces the gas molecules to flow in the direction of the cold surface. In this pressure region, the volumetric pumping speed of the cryo-pump increases above the constant value attained during the free molecular flow region. At liquid helium temperature, the only gaseous vapour pressure is that of helium itself, hydrogen and neon. These gas loads may be removed by auxiliary pumps such as diffusion pumps or cryo-sorption pumps. For cost considerations, many pumps are operated at 20 K (liquid hydrogen temperature). Since cryo-plates possess limited gas handling capacity, they are periodically warmed up to release gases. The cryo-pumps are generally cooled by compressed helium or using a closed loop refrigeration system. With saturation of the surface with condensate, the pumping speed becomes zero and regeneration becomes necessary. Saturation is much faster at high operating pressures. Regeneration can be achieved by evaporation of the condensed gases. Generally, the chamber is heated under vacuum to the highest possible temperature and pumped by the mechanical pump before cooling the cryo-pump in vacuum. The cryo-pumps can also be combined with the cryo-sorption pumps. In such cases, the cold surface is coated with zeolites. The lighter gases such as helium and hydrogen have very low trapping efficiency and their presence is observed in ultra-high vacuum systems. The cryo-pumps find applications also in high vacuum furnaces and in work associated with wind tunnels and rocket testing.

6.2.4.2 Cryo-Sorption Pumps

Gas particles impinging on a cold surface lose much of their incident kinetic energy and remain attached to the surface by weak intermolecular forces. This results in much higher molecular concentration on the surface than in the gas phase. This phenomenon, known as cryo-sorption is physical adsorption under vacuum conditions and low temperatures. In this case, the equilibrium pressure of adsorbed gas is significantly lower than the corresponding saturation pressure for cryo-condensation. Thus, the gas can be retained by adsorption in a sub saturated state at considerably higher temperatures than would be required for condensation. This makes it possible to cryo-pump helium, hydrogen, and neon, which cannot be condensed at higher temperature. Equilibrium pressures in the range of 10^{-7} Pa range can be achieved for helium and hydrogen at temperature of 5 K. Cryo-sorption pumping is limited to few monolayers of gas coverage on the surface.

Molecular sieves (zeolites) or activated charcoal are used as sorbent materials in construction of these pumps. Factors considered for selection of sorbents for the pumps include

- Sorption capacity

- Thermal conductivity
- Ease of refrigeration
- Dusting characteristics
- Explosive hazards
- Cost

Synthetic zeolites which are alumino-silicates of Na, K or Ca are most suitable sorbents for these pumps.

Activated charcoal is a highly porous form of carbon, free from graphite, ash and any hydrocarbon residuals. Hardwoods and nutshells used as sorbents possess a structure of long, interconnected cellulose fibers built of groups of carbon, hydrogen and oxygen atoms arranged to form almost endless chain. Processing expels hydrogen and oxygen but carbon atoms probably retain their chainlike arrangement with some cross-linking but with many unsatisfied valencies. The resulting porosity includes passage of atomic dimensions. The material is cut into small pieces, freed from saw-dust and heated in closed but vented steel chamber at 500 to 700°C until all hydrocarbon is removed and no visible vapour evolution is observed. It is further treated at 600°C in dry vacuum and stored in vacuum. In an alternative method, passage of steam or CO_2 is passed through charcoal heated to 800°C to 1000°C. Activated charcoal Type 208-C is composed of coconut shell, steam activated charcoal and has an average pore diameter of 10 Å.

Molecular sieves are synthetic, dehydrated, crystalline alumino-silicates. Their internal cavities are normally filled with water of crystallization. The unique property of the crystals being that the crystal structure remains unchanged even after the removal of the water of crystallization by heating and does not collapse. The internal cavities are interconnected by pores of uniform diameter. About half the volume of the crystal may consist of cavities that are available for sorption of such gases which are able to penetrate the interconnecting pores. Molecules of unsuitable size and shape are unable to pass through the pores so the zeolite structure acts as a sieve, excluding the larger and more awkwardly shaped molecules while permitting the smaller ones to enter. Molecular sieves are available in 1/8 inch or 1/16 inch cylindrical pellets. The following Table 6.1 gives details of the different types of molecular sieves that are available commercially.

Table 6.1. Commercially available molecular sieves

Molecular sieve type	Composition of alumino-silicate	Approximate pore size (Å)
3A	K	3
4A	Na	4
4A-XW	Na	4
5A	Ca	7
10X	Ca	9
13X	Na	10

Synthetic zeolites possess higher sorption capacity than activated charcoal but the latter has a higher thermal conductivity. Ease of refrigeration is better for charcoal while zeolites offer less dusting. Explosion hazards are relatively less in zeolites but the costs are comparatively higher. 100 gms of molecular sieve type 13X has a surface area of 5.14×10^8 cm^2, and when cooled to 77 K, it has a relative coverage of N_2 as $\theta = 0.1$ at a pressure of 10^{-3} Torr. This corresponds to 3.2×10^{22} N_2 molecules adsorbed, which is approximately the number of molecules in a volume of 1 liter at STP. This illustrates that modest quantities of porous materials can serve as fore-pumps to reduce the pressure from the atmosphere. Design calculations for cryo-sorption pumping have been given by Manes and Grant[37]. Jepsen et al[38] have described the arrangement for which the system was sealed from atmosphere and a finger containing activated charcoal was immersed in liquid nitrogen after which the pressure dropped to 1 Pa.

The cryo-sorption pump consists of a stainless steel container of moderate thermal mass. The sorbent bed has a large surface area against the pump body wall so that the cooling of the low thermal conductivity sorbent is rapid. The gas to be pumped should have a free access to a large area of the sorbent bed. Sorption pumps are provided with heater jackets or rods for activation/degassing. Vanes are provided inside the pump for better thermal conductivity. Good thermal contact between the sorbent and the underlying cooled surface is essential to establish rapid thermal equilibrium. A typical cryo-sorption pump is shown in Fig. 6.30. The cryo-sorption pump producing the ultimate pressure of 10^{-7} Torr in a 2 liter volume has been described by Nair and Vijendran[39].

The cryo-sorption pumps are commonly employed for roughing the vacuum systems. Low-boiling point gases such as Ne, He and H_2 in the atmospheric air are not pumped effectively. This leaves

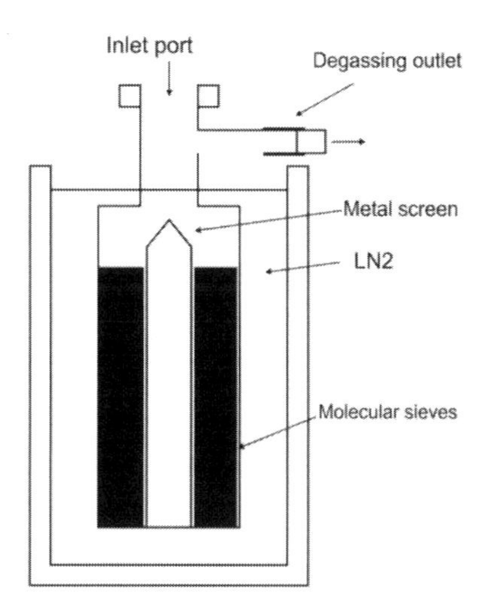

Fig. 6.30. Cryo-sorption pump.

a considerable gas load of noble gases to start with. Initially Ne is flushed out with the bulk of the gas due to viscous drag. With the approach of molecular flow conditions at lower pressures, back diffusion of Ne begins to occur.

Sequential operation of the cryo-sorption pumps is employed to minimize the Ne content in the system. To achieve this, one cryo-sorption pump evacuates the system to about 100 Pa and is isolated. This way, most of Ne is pumped with the viscous flow and has no chance to diffuse back into the system. The other pump then takes over to further evacuate the system. Figure 6.31 shows the pumping action with the sequential operation with two pumps which is more effective.

The limiting pressure with the cryo-sorption pump is established by an appropriate isotherm[40] relating the equilibrium pressure to the surface coverage. A cryo-surface with less than monolayer coverage at 4.2 K will pump all gases to extremely low pressures.

The cryo-sorption pumps can be used for initial evacuation with systems which do not need continuous backing pumps. Thus these pumps are best suited for systems using getter-ion/sputter-ion pumps. The cryo-sorption pumps offer clean vacuum, free from organic contamination and are silent as no vibrations occur. However, these pumps cannot handle large gas loads and do require continuous refrigeration during their normal operation. These pumps cannot be used for pumping mercury vapours.

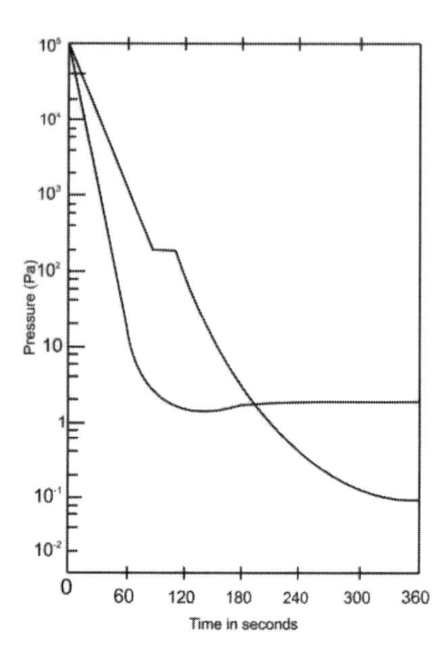

Fig. 6.31. Sequential operation of cryo-sorption pumps.

6.2.4.3 Pumping by Cryo-Trapping

Cryo-trapping involves concurrent pumping of two or more gases by entrainment of gas particles. A condensing gas is used for entrainment to form a mixed condensate. The smaller molecules of the gas to be pumped are trapped in the open lattice of the cryo-deposit of a more abundant species and are buried by subsequent layers. Most bonded gas particles are occluded in the condensate layer leaving only a certain fraction to directly interact with the surface. Thus, the equilibrium pressures are lower than those that can be achieved by cryo-sorption. This type of pumping requires additional gas input into the system which has to be pumped thereby requiring larger forevacuum systems to handle the additional load during regeneration.

In the under-saturated condition, hydrogen can condense on cold surfaces if mixed with argon. The mixed condensation of argon and hydrogen for condensation surface temperatures between 4.2 K and 15 K was investigated by Trendelenburg[41] who suggested that binding of hydrogen in the argon condensate occurs with two different binding energies. His results showed that one of the binding energies was of about 700 cal/mole which is higher than the heat of evaporation of pure hydrogen and less than the adsorption energy of hydrogen on metal surfaces.

6.2.4.4 *Types of Cryo-Pump*

Two types of cryo-pumps are commonly used include the bath type and the refrigerator-cooled type.

In bath cryo-pumps dewar and cryostat designs are used. Cryogenic liquids are stored in dewar vessels. The pumping surfaces facing the vacuum vessel are directly cooled from their rear side by cryogenic liquids boiling at ambient pressure. Thus, temperatures of 77.3 K, 27.1 K, 20.3 K and 4.2 K corresponding to liquid nitrogen, liquid neon, liquid hydrogen and liquid helium respectively can be attained. The pumping surface needs to be always shielded from thermal radiation by a liquid nitrogen-cooled stage. Longer holding times with continuous operation can be obtained by using an automatic device for refilling. Further fractionation of the gases to be pumped can be achieved by using an additional baffle, which is cooled with liquid helium, to shield the 4 K liquid helium cryo-

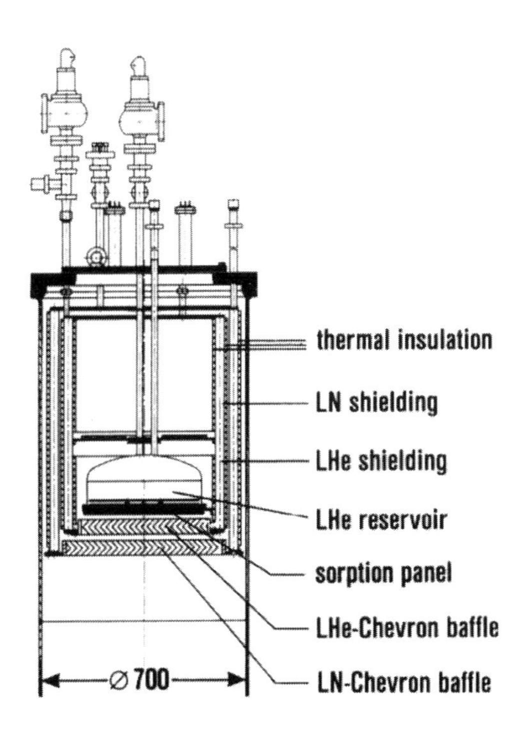

Fig. 6.32. Bath type cryo-pump *(Courtesy: C. Day).*

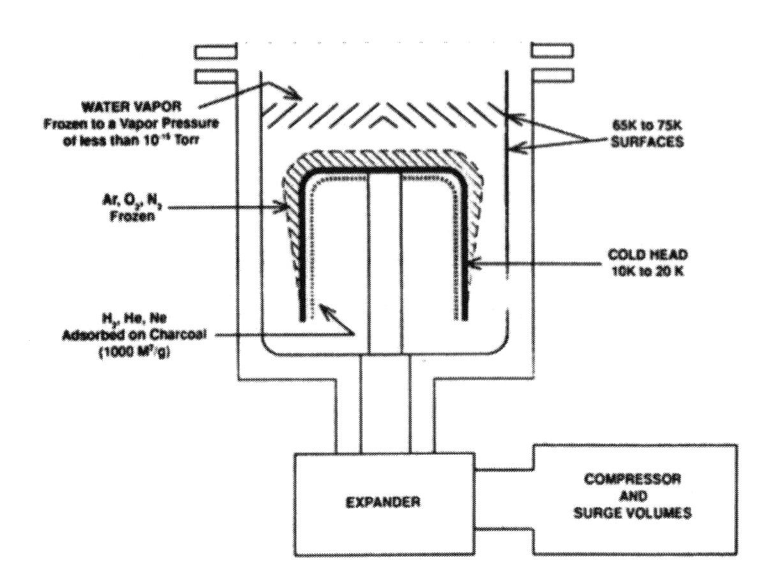

Fig. 6.33. Two-stage refrigerator cooled cryo-pump *(Courtesy: C. Day).*

sorption panel. Figure 6.32 illustrates the bath type cryo-pump. Figure 6.33 shows a two-stage refrigerator cooled cryo-pump.

Advantages of cryo-pumps include absence of moving parts except those in the refrigeration system, absence of contamination by oil, grease, compact size and the requirement of backing pump only during starting and regeneration.

References

1. W. Becker, Vak. Tech. **7**, 149 (1958).
2. H. Enosawa, C. Urano, T. Kawashima, and M. Yamamoto, J. Vac. Sci. Technol. A, **8**, 2768 (1990).
3. B. L. Cho, S. T. Lee and S. M. Chung, J. Vac. Sci. Technol. A, **13**, 2228 (1995).
4. NASA Tech Brief Number GSC-16695-1 September 2013; 19
5. B. D. Power and D. J. Crawley, Vacuum **4**, 415 (1954).
6. M. N. Hablanian and H. A. Steiherz, 1961 Vacuum Symp. Trans., 333 (1962).
7. D. Alpert, Rev. Sci. Instrum. **24**, 1004 (1953).
8. D. J. Santeler, J. Vac. Sci. Technol. **8**, 299 (1971).
9. P. della Porta, T. Giorgi, S. Origlio and F. Ricca, Proc. 8th Nat. Symp. AVS, London, 229 (1962).
10. B. Kindl, Supplem. Al Nuovo Cemento **8**, 646 (1963).
11. Y. Suetsugu, K. Shibata and M. Shirai, Nucl. Instrum. Methods in Phys. Res. Sect. A **597**, (2–3), 153 (2008).
12. C. Benvenuti and P. Chiggiato, Vacuum **44**, 511 (1993).
13. C. Benvenuti, P. Chiggiato, F. Cicoira and Y. Aminot, J. Vac. Sci. Technol. A **16**, 148 (1998).

14. R. G. Herb and R. H. Davis. Procs, First Nat. AVS Symp.1954, W M Welch Manufacturing Company, 40 (1955).
15. K. A. Warren, D. R. Denison, and D. G. Bills, Rev. Sci. Instrum. **38**, 1019 (1967).
16. J. Harra and T. W. Snouse, J. Vac. Sci. Technol. **9**, 552 (1972).
17. M. V. Kuznetsov, A S Nasarov and G. F. Ivanovski, J. Vac. Sci. Technol. **6**, 34 (1969).
18. R. E. Clausing, Report ORNL-3481 UC-25- Metals Ceramic and Materials-TID 4500 (1964).
19. A. K. Gupta and J. H. Leck, Vacuum **25**, 362 (1975).
20. G. M. McCracken, Vacuum **15**, 433 (1965).
21. F. Prevot and Z. Sledziewski, J. Vac. Sci. Technol. **9**, 49 (1972).
22. T. Pauly, R. D Welton and R. G. Herb, 1960 Vac. Symp. Trans., 51 (1961).
23. R. G. Herb, Revista Brasileira de Física, Vol. 2, N." 1, (1972).
24. P. A. Redhead, E V Kornelsen and J P Hobson, Can. J. Phys. **40**, 1814 (1962).
25. H. Huber and M. Warnecke, Vide **74**, 84 (1958).
26. R. A. Douglas, J. Zabritski, and R. G. Herb, Rev. Sci. Instrum. **36**, 1 (1965).
27. P. K. Naik and R. G. Herb, J. Vac. Sci. Technol. **5**, 42 (1968).
28. P. K. Naik and S. L. Verma, Proc. 6th Internl. Vac. Congress 1974, Jap. J. Appl.Phys. Suppl. 2 Part 1 (1974).
29. P. K. Naik and S. L. Verma, J. Vac. Sci.Technol. A **14**, 734 (1977).
30. S. L. Rutherford, S. L. Mercer and R. L. Jepsen, Proc. 7th Nat. AVS Symp., 1960, Pergamon Press, Inc., New York, 380, (1961).
31. R. L. Jepsen, A. B. Francis, S. L. Rutherford, B. E. Kietzman, Proc. 7th Nat. AVS Symp. 1960, Pergamon Press, Inc., New York, (1961).
32. L. T. Lamont , J. Vac. Sci. Technol. **6**, 47 (1969).
33. T. Tom and B. D. James, J. Vac. Sci. Technol. **6**, 304 (1969).
34. W. M. Brubaker, Proc. 6th Nat. AVS Symp. 1959, Pergamon Press, Inc., New York, 302, (1960).
35. J. M. Laurent, and O. Gröbner, Distributed Sputter-Ion Pumps for use in Low Magnetic Fields, Particle Accelerator Conference, San Francisco (1979).
36. C. Day, Basics and Applications of Cryo-Pumps, Proc. CERN Accelerator School on Vacuum, Platja d'Aro, Spain, 241 May 2006.
37. M. Manes and R. J. Grant, Trans. 10th American Vac. Symp. 122, Macmillan New York, (1963).
38. R. L. Jepsen, S. L. Mercer and M. J. Callaghan, Rev. Sci. Instrum. **30**, 377 (1959).
39. C. V. G. Nair and P. Vijendran, Jap. J. Appl. Phys. **13**, 93 (1974).
40. J.P. Hobson, J. Phys. Chem. **73**, 2720 (1969).
41. E. A. Trendelenburg, Vacuum **17**, 495 (1967).

7

System Design

The material of construction of vacuum systems is generally hard glass or stainless steel. These materials, after adequate cleaning and processing offer the lowest outgassing rates. Glass systems are most suitable for laboratory experimental work in smaller (few liters capacity) volumes. However, one needs to employ glass-to-metal seals for providing electrical connections. Stainless steel systems need to utilize flanged or union coupling joints. Most of the permanent joints in the metal system are generally inert gas arc welded or vacuum brazed. The hardware in the system is selected depending upon the degree of vacuum. Elastomer gaskets can be used for flanged joints and valves in high vacuum systems. However, bakeable hardware with soft metal gaskets is required for UHV systems.

It is important that the conductance C of the pipe or the component connecting rest of the system to the pump should have conductance equal to or higher that the speed S of the pump to obtain the maximum possible effective speed S_e, as can be observed from the equation

$$S_e = \frac{SC}{(S+C)}$$

It can be seen that for $C \gg S$,

$$S_e = S$$

It is therefore necessary to use a wide-bore short length connecting pipe to maximize the effective speed.

The first stage of pumping involves evacuation from the atmospheric pressure to about 10 Pa. The load due to gassing from the walls can be considered only in case of large surface area exposed to the pumping system or if the pump speed is much lower compared to the system volume.

7.1 Selection of Pumps

Selection of pumps is primarily governed by the application, the process to be undertaken in vacuum, the pressure range of operation, the quality of vacuum required, the volume, the gassing loads, the pump-down time requirement, the pumping speed.

Generally, the rotary vane and rotary piston pumps with traps are suitable for evacuating a few liters volume from atmospheric pressure down to pressures below 10^{-1} Pa, as these pumps are capable of working continuously at low pressures. If the volume to be evacuated contains significant amount of moisture, as is common in humid conditions, it is advisable to run the rotary pump in the gas–ballast mode to start with in order to avoid contamination of the pump oil with water which can cause a back pressure that would hamper pumping.

In case the pumping is required at pressures higher than 10^4 Pa over long periods, jet pumps, water ring pumps are preferable. If a large amount of gas is required to be handled at pressures below 4×10^3 Pa, a Roots pump, backed by a single-stage or two-stage rotary pump, works satisfactorily. Two-stage Roots pumps with two-stage rotary backing pumps can offer high pumping speed in the pressure range between 1 Pa and 10^{-2} Pa. This combination works satisfactorily for unattended operation. Alternatively, a vapour ejector pump can be used in this pressure range

Vapour boosters are commonly used in the 10^{-2} to 10 Pa range where primary pump combinations are often at their limit and ordinary diffusion pumps exhibit instability. Vapour booster pumps are capable of pumping contaminated systems and processes with high gas loads of hydrogen; hence they are particularly suited for use in metallurgical and chemical process applications. These pumps offer very large pumping speed at high operating pressures and can handle very high throughput at operating pressures.

The diaphragm pumps can be employed as backing pumps for turbo-molecular pumps. These pumps can be used for vacuum filtration/distillation/drying.

During the stage of pump-down from 10 Pa to about 10^{-2} Pa, significant gassing occurs. High vacuum pumps such as diffusion pump or TMP or cryo-condensation pumps are utilized in this pressure range with rotary mechanical pump as the baking pump or a fore-pump. It is important that the speed of the backing pump is sufficient to maintain a pressure that is required by the high vacuum pump at its exhaust port for the gas loads handled.

For applications in the pressure range between about 1×10^{-4} Pa and 1×10^{-1} Pa, with a few liters volume to be evacuated, a system with oil-diffusion pump, backed by an oil-sealed rotary pump should be satisfactory. For the diffusion pump to perform satisfactorily, it is necessary to limit the fore-pressure. Refrigerated traps between the pumps and between the diffusion pump and the chamber are required in order to minimize possible contamination by pump fluid vapours, in addition to a baffle over the diffusion pump. A roughing line that bypasses the diffusion pump is required to be incorporated to pre-evacuate the chamber by the roughing pump to prevent contamination of the diffusion pump fluid during the initial pump-down at high pressure. The configuration of the vacuum system can be similar for larger volumes except that the higher speed capacity diffusion pump and the backing rotary pump should be utilized. Running the rotary pump in the gas–ballast mode to start with becomes necessary for large volume systems, particularly if the chamber is filled with atmospheric air in the previous operation and has remained unused for longer period. Gas ballasting may not be necessary if the chamber is filled with dry nitrogen or argon at the end of the previous operation. A typical vacuum system using a diffusion pump and a rotary mechanical pump is shown schematically in Fig. 7.1. Specially designed diffusion pumps that combine the characteristics of the diffusion pump such as capability of achieving low ultimate pressures, high pumping speeds in the high vacuum region with the characteristics of a vapour ejector pump such as capability of handling high throughput in pressure range between 1 Pa and 10^{-4} Pa are effectively used.

Starting of the diffusion and sputter-ion pumps require that the vacuum chamber is pre-evacuated to pressures of about 10 Pa to 10^{-1} Pa which can be achieved by rotary mechanical pumps. Operation of the diffusion pump demands continuous backing by the rotary vacuum pump. However, the operation of SIP does not need the backing pump after it is started.

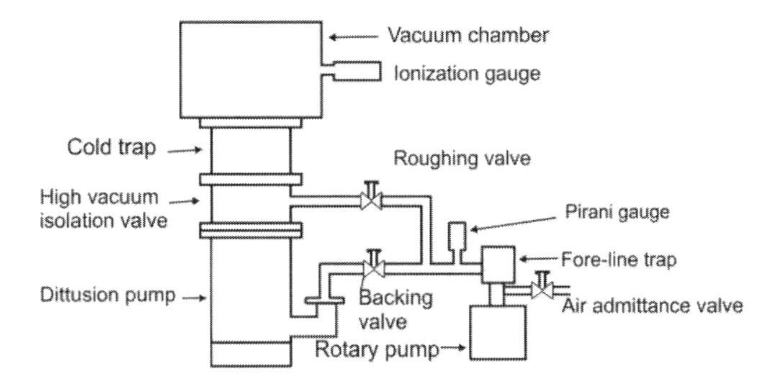

Fig. 7.1. Schematic of a typical diffusion pumped vacuum system.

For applications requiring a clean organic contamination-free vacuum, it is necessary to employ a cryo-pumped system or an ion pumped system. As in the case of most high vacuum pumps, fore-pressure needs to be provided by a mechanical pump before the cryo-pump is started. The mechanical pump can also be used during the regeneration of the cryo-pump while the condensed gas is exhausted by heating. For applications that demand a clean organic contamination-free vacuum, it is necessary to employ a cryo-sorption pumped system.

Systems using combinations of sputter-ion pumps, getter-ion pumps, titanium sublimation pumps non-evaporable getter pumps, turbo molecular pumps, cryo-sorption pumpd are most commonly used for experimental work in UHV range, surface analytical equipment and particle accelerators.

7.2 Pumping Process and Pumpdown

The gas influx resulting from a particular process that takes place in vacuum needs to be handled in such a manner that the pressure near the pump inlet does not keep on rising continuously and an equilibrium pressure within the operating pressure range of the pump is required to be maintained during the process. The effective pumping speed is determined by the combination of the actual speed of the pump at a particular pressure and the conductance of the component such as baffle, trap, valve, elbow that is interposed between the pump and the chamber. Thus the combination of the pump and the interposed component should provide a speed sufficient to pump down the chamber in short time and maintain equilibrium pressure well within the upper operating limit of the pump.

As discussed earlier, the basic equation (2.35) for pumping is given by

$$V\frac{dP}{dt} = -PS + Q$$

where V is the volume of the chamber being pumped; P is the pressure of gas in the chamber at time t; S is the pumping speed of the pump; Q is the rate of the amount of gas influx (throughput); P_u is the ultimate pressure, the lowest steady pressure achieved upon pumping is expressed in equation (2.36)

$$P_u = \frac{Q}{S}$$

If we further consider that there are no leaks the throughput Q is solely due to the gassing from sources other than leaks and is independent of pressure

$$-V\frac{dP}{dt} = S\left(P - P_u\right) \tag{7.1}$$

The effective pumping speed S_e can be considered as

$$S_e = -\left(\frac{1}{P}\right)V\frac{dP}{dt}$$

Thus, we have

$$S_e = S\left(1 - \frac{P_u}{P}\right)$$

Thus, the effective speed keeps on falling as P becomes comparable to P_u and $S_e = 0$ at $P = P_u$.

However, the actual pumping speed S that is still available is given by

$$S = \frac{Q}{P_u}$$

Considering Q and S constant, we get,

$$P - P_u = \left(P_o - P_u\right)e^{\frac{-St}{V}}$$

where P_0 is the pressure at time $t = 0$. Also as P_0 and $P \gg P_u$

$$P = P_o e^{\frac{-St}{V}} \tag{7.2}$$

$$P_o = P e^{\frac{St}{V}}$$

$$\frac{P_o}{P} = e^{\frac{St}{V}}$$

$$\frac{St}{V} = \ln\left(\frac{P_o}{P}\right) \quad \text{and}$$

$$t = \ln\left(\frac{P_o}{P}\right) \times \frac{V}{S} \tag{7.3}$$

$$t = \frac{V}{S} \times \log_{10}\left(\frac{P_o}{P}\right) \times 2.30$$

$$= \tau \times \log_{10}\left(\frac{P_o}{P}\right) \times 2.30$$

With $P \gg P_u$, the term V/S decides the performance of the pumping system and is called time constant τ. Thus, the pumpdown time from the initial pressure P_o at $t = 0$ to a lower pressure P at time t can be calculated if V and S are known. Practically, the pumpdown takes a much longer time, primarily due to slower removal of water vapour in the vacuum system. This effect is observed particularly at pressure range below about 10^{-1} Pa.

Redhead[1] has shown that the gas throughput from other sources such as thermal outgassing, heated components, diffusion from the walls, charged particle/radiation induced desorption has to be considered for determining the ultimate pressure in UHV and XHV systems. He has established that the maximum pumping speed is independent of volume and surface area of the system if the combined gas loads from the outgassing of the walls and the diffusion from the walls are much higher than the sum of the combined gas loads due to heated surfaces and the charged particle/radiation induced desorption.

Lewin[2] has discussed pumping of a distributed volume. A tube of uniform cross section diameter of length L and closed at one end is considered with the open end connected to an ideal leak-tight pump of speed S that does not release gases. The gases to be handled by the pump are the gas in the volume and the outgassing from the walls.

As shown in Fig. 7.2, a tube element of length dx is considered. Then the change of flow between cross sections x and $x+dx$ will be equal to the difference between the gassing rate and the rate of removal of the volume gas. x is taken as the distance from the open end of the tube and P is the pressure of the gas. The volume per unit length v is taken as V/L where V is the volume. If

$$q = \frac{Q_w}{L}$$

where q is the specific gassing rate and Q_w is the total gassing rate.

It is assumed that initially a uniform pressure P_0 exists in the tube and the flow into the closed end is zero. In the equilibrium condition after pumpdown with a constant gassing rate q, it is established that P decreases parabolically with x from P_{max} at $x = L$ to

$$P = \frac{qL}{S} \text{ at the pump port}$$

7.2.1 Pumpdown Time

Consider a stainless steel cubic chamber with each side of 70 cm. The chamber is evacuated using a diffusion pump and a mechanical rotary pump as shown in Fig. 7.3. A baffle valve (intermediate component) is interposed between the chamber and the diffusion pump. A roughing tube of 3.8 cm diameter and 70 cm length is used between the baffle valve and the rotary pump. The specific outgassing rate from the walls (q) is 1.7×10^{-6} Pa · m · s^{-1}. The ultimate pressure (P_u) is 10^{-4} Pa in the chamber. An equilibrium pressure (P_e) of 10^{-1} Pa is achieved upon bleeding nitrogen in the

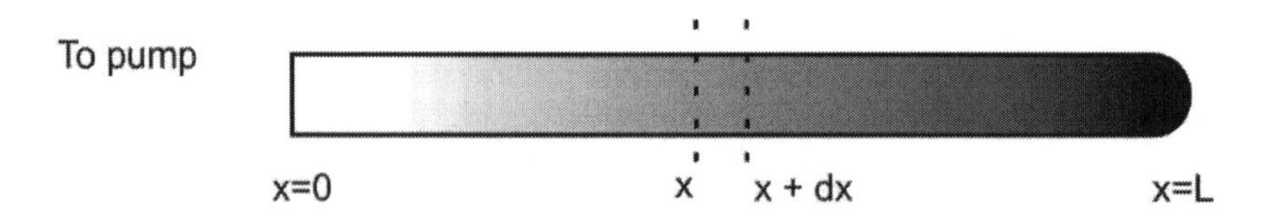

Fig. 7.2. Distributed volume evacuation.

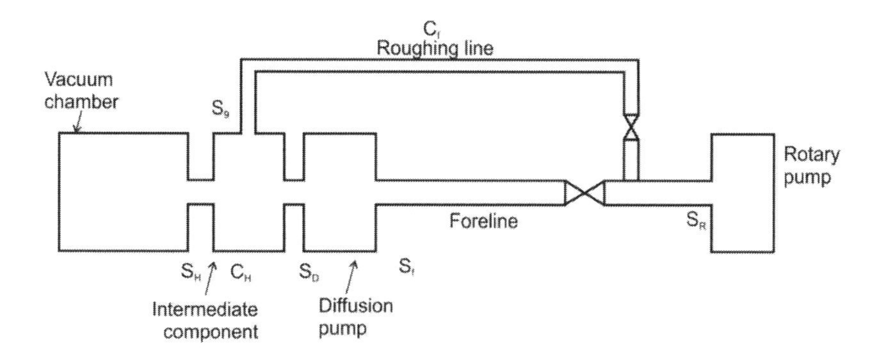

Fig. 7.3. Typical vacuum system with interconnecting line.

chamber. The limiting fore pressure P_f at which the diffusion pump operates is 20 Pa. The inlet pressure P_{in} for starting the diffusion pump using the rotary pump is 10 Pa. It is required to decide speeds of the diffusion pump and the rotary pump.

Surface area A of the chamber is 2.94 m^2; volume V of the chamber is 0.343 m^3; specific outgassing rate (q): 1.7×10^{-6} $Pa{\cdot}m{\cdot}s^{-1}$ total outgassing rate (Q_g), given by $q \times A$ is 5×10^{-6} $Pa \cdot m^3 \cdot s^{-1}$.

Ultimate pressure (P_u) of 10^{-4} Pa in the chamber is achieved resulting from the outgassing rate and the pumping speed available at the chamber and is given by

$$P_u = \frac{Q_g}{S_h}$$

where S_h is the pumping speed available at the chamber.

$$\text{Thus } S_h = \frac{Q_g}{P_e} = 5 \times 10^{-2}\,\text{m}^3 \cdot \text{s}^{-1}$$

Assuming the pumping speed of the diffusion pump S_d equals the conductance (C_h) of the intermediate component/baffle valve,

$$S_d = C_h$$

$$\text{and} \frac{1}{S_h} = \frac{1}{C_h} + \frac{1}{S_d}$$

$$S_d = 2S_h = (10^{-1})\,\text{m}^3{\cdot}\text{s}^{-1}$$

If a diffusion pump of speed (10^{-1}) $m^3 \cdot s^{-1}$ is used along with an intermediate component of conductance 10^{-1} $m^3 \cdot s^{-1}$, the effective speed S_h at the intermediate component inlet will be 0.05 $m^3 \cdot s^{-1}$.

Equilibrium pressure (P_e) of 10^{-1} Pa is achieved upon bleeding nitrogen in the chamber. The throughput Q handled will be given by

$$Q = P_e \times S_h = 0.005 \text{ Pa·m}^3 \cdot \text{s}^{-1}$$

The throughput Q is required to be handled by the forevacuum side with the limiting fore pressure P_f of 20 Pa and forepumping speed S_f.

Thus

$$Q = P_f \times S_f$$

$$0.005 = 20 \times S_f$$

$$\text{and} \quad S_f = 2.5 \times 10^{-4} \text{ m}^3 \cdot \text{s}^{-1} \tag{7.4}$$

Although with this speed the foreline can offer adequate backing for the diffusion pump, the speed is too small to evacuate a chamber of 343 liters from the atmospheric pressure to a pressure of 10 Pa required to start the diffusion pump as it will result into a very long duration of pumpdown time. Therefore a much higher roughing line speed is necessary. If a desired pumpdown time for roughing from the atmospheric pressure of 10^5 Pa to the inlet pressure P_{in} of 10 Pa for starting the diffusion pump is 20 minutes, the effective roughing speed S_g required at the intermediate component can be decided by using the equation (7.3)

$$t = \ln\left(\frac{P_0}{P}\right) \times \left(\frac{V}{S}\right)$$

Thus,

$$S_g = \ln\left(\frac{P_0}{P}\right) \times \left(\frac{V}{t}\right) = \frac{0.343}{1200} \ln\left(\frac{10^5}{10}\right) = 2.62 \times 10^{-3} \text{ m}^3 \cdot \text{s}^{-1}$$

As a roughing tube is used between the chamber and the rotary pump, its conductance C_f will be given by the equation

$$C_f = \frac{188 \times 10^3 D^4 P'}{133.28 L} \text{ m}^3 \cdot \text{s}^{-1} \text{ for } N_2 \text{ at 293 K for viscous flows}$$

For $D = 3.8$ cm and $L = 70$ cm,

$$C_f = 4.2 \cdot 10^{-3} P' \text{m}^3 \cdot \text{s}^{-1} \tag{7.4}$$

where P' is the mean pressure in Pa.

Using equation (7.4), the values of C_f are found to vary between 4.2×10^2 m$^3 \cdot$ s^{-1} and 5.2×10^{-2} m$^3 \cdot$ s^{-1} for pressures between 10^5 Pa and 10^2 Pa as the pumpdown occurs in the viscous flow range.

$$\text{As} \quad \frac{1}{S_g} = \frac{1}{C_f} + \frac{1}{S_r}$$

where S_r is the speed of the rotary pump and S_g is the effective speed available at the junction of the foreline and the baffle valve as shown in Fig. 7.3

As $C_f \gg S_r$, $S_r = S_g$ over most of the pressure range. This gives $S_r = 2.62 \times 10^{-3}$ m$^3 \cdot$ s^{-1}.

Thus a backing pump of speed 3×10^{-3} m$^3 \cdot$ s^{-1} can be selected.

The effective pumping speed S_h available at the HV line port near the vacuum chamber is 5×10^{-2} m$^3 \cdot$ s^{-1}.

The HV pumpdown time t_1 from 5 Pa down to 5×10^{-4} Pa is given by

$$t_1 = \frac{V}{S_h}\left[\ln\left(\frac{P_1}{P_2}\right)\right] = 63.18 \text{ s}$$

However, in actual practice, the pumpdown takes a much longer time, primarily due to slower desorption of water vapour condensed on the walls of the vacuum chamber and the connecting HV line. This effect is observed particularly at pressure range below about 10^{-1} Pa.

It can be seen from the discussion above that the time required from roughing the chamber from the atmospheric pressure to 5 Pa is much longer than the time taken for further evacuation by the diffusion pump.

7.3 Symbolic Diagrams

Line diagrams of vacuum systems can be effectively shown by using symbols for various vacuum components as illustrated below in Figs. 7.4 (a), (b), (c) and (d). These figures are reproduced with permission from 'Graphic Symbols in Vacuum Technology', Journal of Vacuum

Science and Technology **4**, 139 (1967), copyright American Vacuum Society.

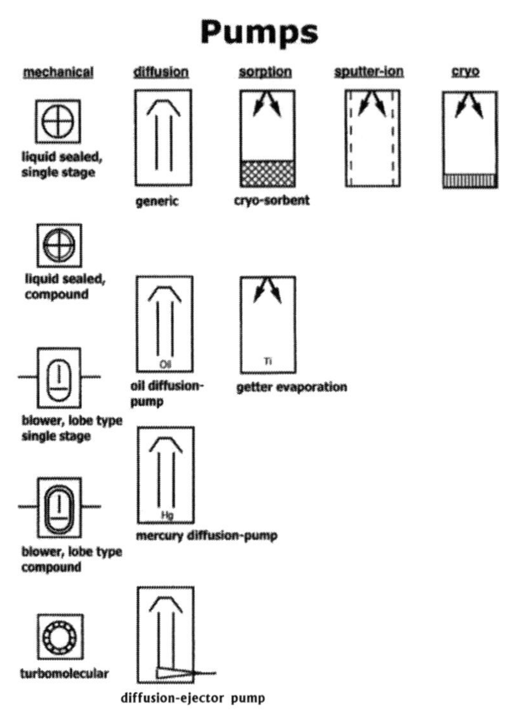

Fig. 7.4 (a). Symbolic diagrams for vacuum pumps (*Courtesy: American Vacuum*

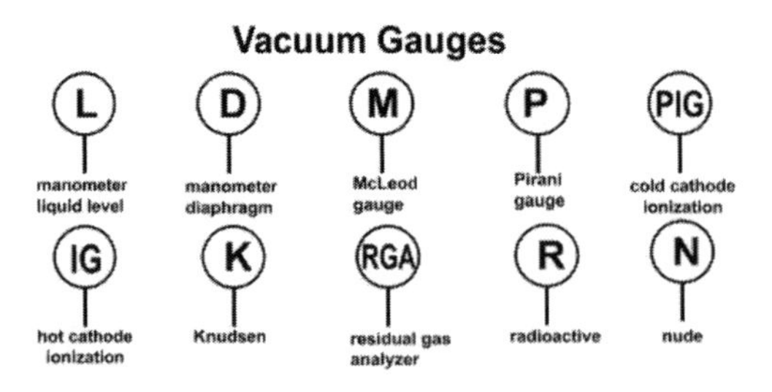

Fig. 7.4 (b): Symbolic diagrams for vacuum gauges (*Courtesy: American Vacuum Society*).

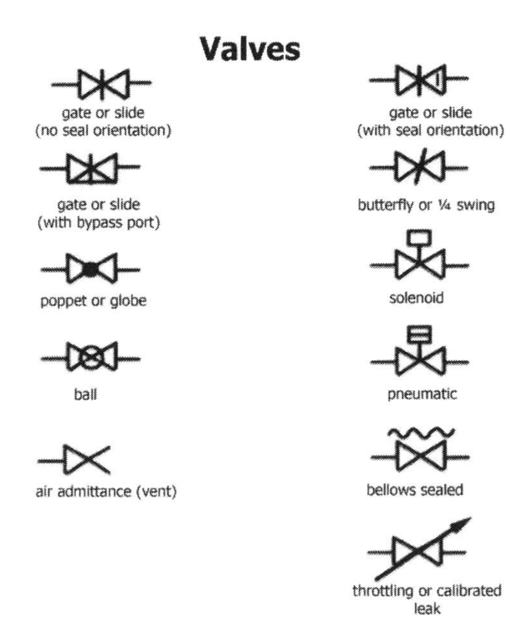

Fig. 7.4 (c): Symbolic diagrams for vacuum valves (*Courtesy: American Vacuum Society*).

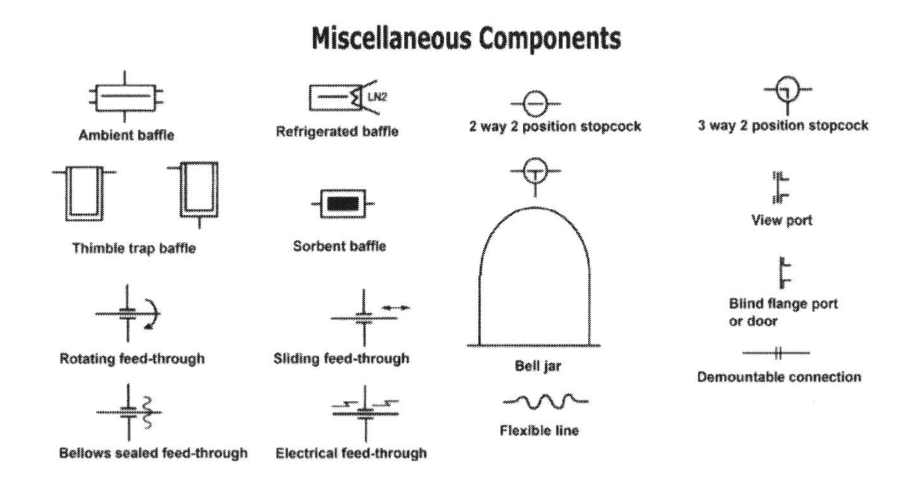

Fig. 7.4 (d): Symbolic diagrams for miscellaneous vacuum components (*Courtesy: American Vacuum Society*).

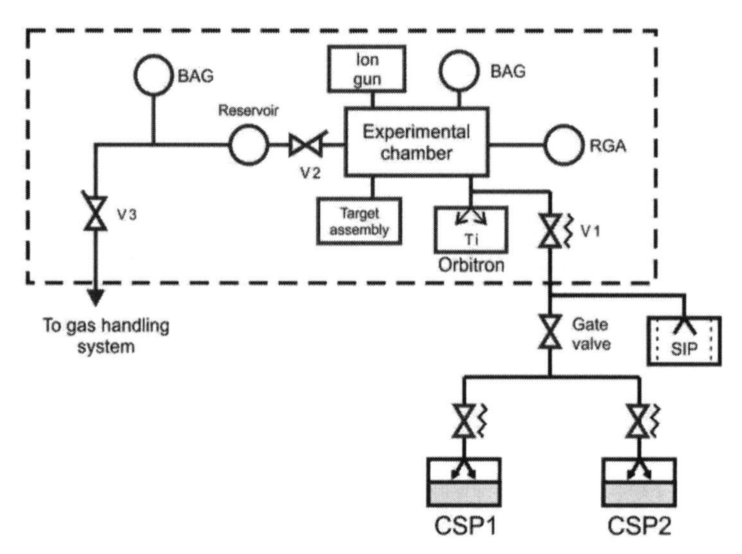

Fig. 7.5. Symbolic diagram of UHV system (*Reprinted from Applied Physics December 1980, Volume 23, Issue 4, pp. 373–380 –Entrapment of inert gas ions near molybdenum surface – P.K. Naik, V.G. Kagal, S.L. Verma, S.P. Mhaskar © by Springer-Verlag 1980 with permission of Springer*).

Figure 7.5 shows symbolic diagram of the UHV system that I used for my work[3].

References

1. P. A. Redhead, "Extreme High Vacuum", Proceedings of the CERN Acceleration School, Snekersten, Denmark, CERN report, edited by S. Turner, 213–227 (1999).
2. G. Lewin, "Fundamentals of Vacuum Science and Technology", McGraw Hill Book Company, New York (1965).
3. P. K. Naik, V. G. Kagal, S. L. Verma and S. P. Mhaskar, Appl. Phys. **23** (4), 373 (1980).

8

Vacuum Measurement Methods

8.1 Pumping Speed Measurements

In the methods for measurement of pumping speed, the gas throughput Q is measured or calculated and its ratio with the pressure P_p near the pump under equilibrium conditions gives the pumping speed. The equilibrium conditions demand that the pressures in the vacuum chamber remain steady for a given magnitude of throughput of the gas. The throughput is measured by the rate of rise of a low vapour pressure fluid in a manometer or a burette of the known cross-section.

$$S = \frac{Q}{P_p} \tag{8.1}$$

In one method, known as the constant pressure method, the leak rate is determined by measuring the time t required for the displaced fluid to change its level through arbitrary distance h in the displacement tube. Referring to Fig. 8.1, if V_0 and B are the initial volume and the initial pressure of the gas filling measuring device; and $(V_0 - V)$ and $(B - P_h)$ are the corresponding values of the gas at time t, then

$$Q = \frac{\left[BV_o - (B - P_h)(V_o - V)\right]}{t} = \frac{\left[BV + P_h(V_o - V)\right]}{t} \tag{8.2}$$

where P_h is pressure due to the head of the fluid and V is the volume displaced at time t.

If the device is designed in such a way that

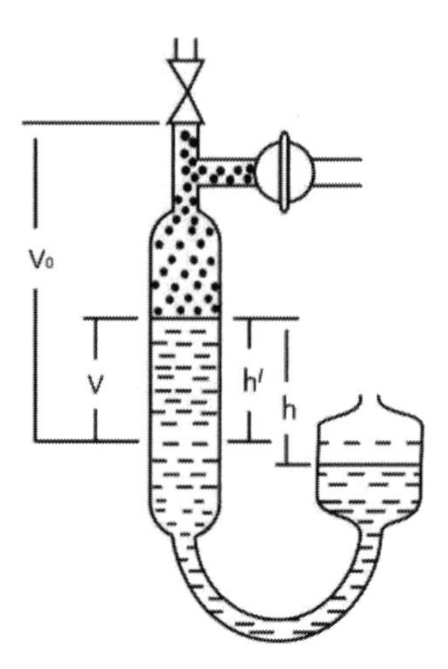

Fig. 8.1. Flow measuring device (*Reprinted with permission from Sets of Tentative Standards, Journal of Vacuum Science and Technology. Copyright American Vacuum Society*).

$$P_h\left(V_0 - V\right) \ll BV$$

then we have

$$Q = \frac{BV}{t}$$

In case the condition $P_h(V_0 - V) \ll$ BV cannot be fulfilled, then equation (8.2) may be employed to use a scale on the device that directly reads the change in the quantity. In the case of smaller values of Q, a small bore tube or pipette with a slug of fluid as shown in Fig. 8.2 *a* may be used.

For flow rates higher than about 7×10^{-1} Pa · m^3 · s^{-1} up to about 7 Pa · m^3 · s^{-1}, a vertical displacement device as shown in Fig. 8.2 *b* can be used.

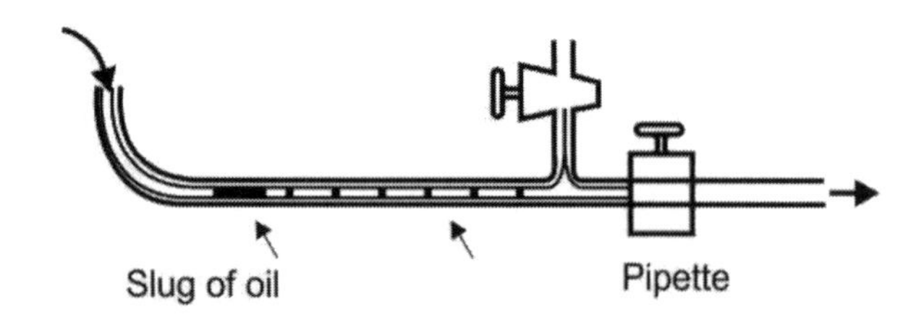

Fig. 8.2a. Apparatus for measurement of gas throughput (*Reprinted with permission from Sets of Tentative Standards, Journal of Vacuum Science and Technology. Copyright American Vacuum Society*).

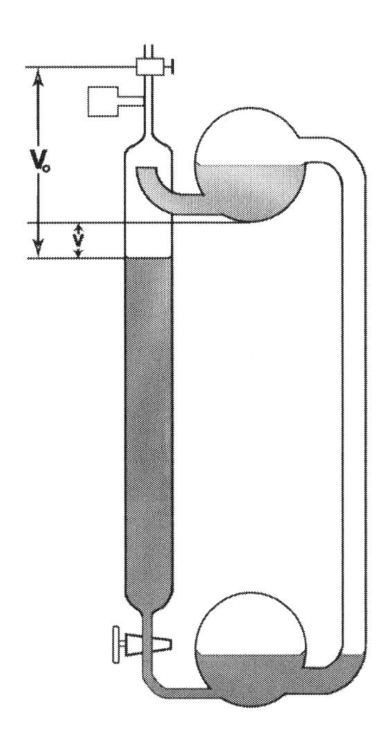

Fig. 8.2b. Vertical displacement device for measurement of gas throughput *(Reprinted with permission from Sets of Tentative Standards, Journal of Vacuum Science and Technology. Copyright American Vacuum Society).*

The other method, known as two-gauge-conductance method, for flow rates less than about 7×10^{-1} to ^{7} Pa · m^{3} · s^{-1} uses an orifice or tube of known conductance C_0 interposed between the source of throughput Q and the pump. The arrangement is illustrated in Fig. 8.3. By adjusting the leak valve, a desired throughput of gas can be achieved that maintains equilibrium pressures on both sides of the conductance with steady pumping. If P_1 is the pressure measured by gauge G_1 on the gas entry side of the orifice and P_2 is the pressure on the other side of the conductance, measured by the gauge G_2 and P_p is the pressure in the gauge G_3 near the pump inlet the pumping speed S is given by

$$Q = C_0 \left(P_1 - P_2\right) = P_p S \tag{8.3}$$

$$\text{If } P_1 \gg P_2 \quad \text{and} \quad P_p \approx P_2$$

$$C_o P_1 = P_p S$$

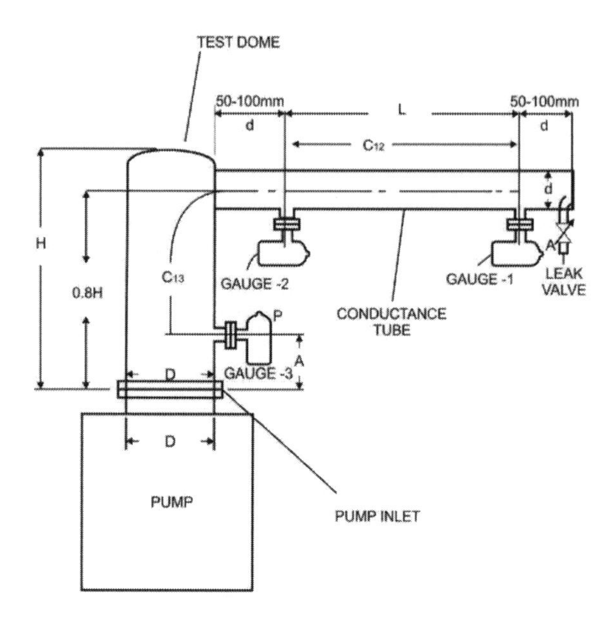

Fig. 8.3. Schematic diagram of the conductance method for pumping speed measurements (*Reprinted with permission from Sets of Tentative Standards, Journal of Vacuum Science and Technology. Copyright American Vacuum Society*).

$$S = C_0 \frac{P_1}{P_p} \tag{8.4}$$

In both these methods, the gas inlet in the vacuum system is required to be directed away from the mouth of the pump to avoid the 'beaming effect'[1]. Alternatively, a standard calibrated gas leak can be used.

Another method involves measurement of pressure during pumpdown of a chamber. The pressure P changes based on

$$P = P_0 e^{-\frac{St}{V}} + P_u \tag{8.5}$$

where P_0 is the initial pressure at time t = 0; P_u is the ultimate pressure; S is the pumping speed; V is the volume of the chamber. We get

$$\log\left[\frac{P_0}{(P-P_u)}\right] = \frac{0.434St}{V} \tag{8.6}$$

For $P \gg P_0$

$$\log\left[\frac{P_0}{P}\right] = \frac{0.434St}{V} \tag{8.7}$$

and

$$S = \frac{-V}{P}\left(\frac{dP}{dt}\right) \tag{8.8}$$

8.2 Calibration of Gauges

The methods for the calibration of ionization gauges include

- The gauge operated at the normal voltages and electron emission current can be directly calibrated against a McLeod gauge of known accuracy at number of pressures measurable by both gauges. The range of calibration is limited to 10^{-3} Pa to 10^{-1} Pa. Calibration at lower pressures can be derived by extrapolation and assumes the response of the gauge to be linear to its lower limit.
- The gauge can be calibrated against the rate of flow through a standard orifice of known conductance.

Lewin[2] has explained that a bakeable diaphragm gauge can be calibrated with the McLeod gauge. Further, a high pressure ion gauge, calibrated with the diaphragm gauge can be used to calibrate a BA gauge. Dushman and Found[3] checked the linearity of the BA-gauge by monitoring the rate of rise of pressure in an evacuated chamber upon leaking the gas from a gas filled reservoir into the chamber through a small conductance. For a constant throughput Q_L of the gas into the chamber of volume V, the pressure P in the chamber would rise linearly with the pressure as in absence of pumping

$$\frac{dp}{dt} = \frac{Q_L}{V}$$

Alpert and Buritz[4] added another chamber further connected by an additional conductance. In this case, the pressure would rise quadratically with time.

References

1. P. Causing, Z. Physik **66**, 471 (1930).
2. G. Lewin, "Fundamentals of Vacuum Science and Technology", McGraw-Hill Book Company (1965).
3. S. Dushman and C. G. Found, Phys. Rev. **17** (1921).
4. D. Alpert, R. S. Buritz, J. Appl. Phys. **25**, 202 (1954).

9

Vacuum Materials, Hardware, Fabrication Techniques, Cleaning Processes and Surface Treatment

The selection of materials and components in vacuum practice is primarily governed by the applications and the pressure range. The glass systems can be conveniently used for experimental studies involving small volumes. Lower outgassing rates, higher chemical stability, excellent electrical resistivity, high thermal shock resistance and lower coefficient of thermal expansion are important characteristics that makes glass suitable for UHV applications. Limitations of glass include its fragility, inability for bakeout at temperatures higher than 450°C, deterioration of insulating properties at high temperatures and inability to construct large (>10 liters) volume chambers. Most UHV systems for experimental studies involving interaction with clean solid surface need to use bakeable components. Stainless steel Type 304 is the most suitable material of construction. Aluminium alloy Al 6061-T6, OFHC copper are also are used for some applications. Moybdenum disulphide (MoS_2) powder with particle sizes in the range 1 to 100 μm is suitable as a lubricant in vacuum environment. Hilton and Fleischauer[1] have discussed the lubricants for high vacuum applications.

9.1 Couplings and Flanged Joints

Figure 9.1 shows a typical demountable vacuum coupling with elastomer O-ring gaskets. Two arms of the metal vacuum system can be coupled using such arrangement. Such couplings are suitable for

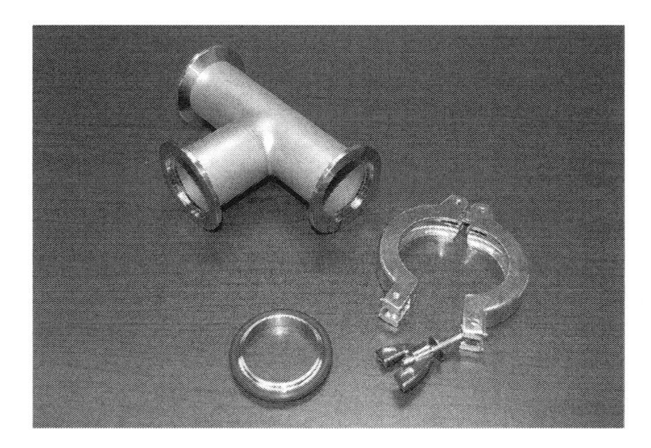

Fig. 9.1. Coupling with elastomer gasket having o-ring retainer (*CC BY-SA 3.0 – Kkmurray K (Wikimedia Commons)- https://commons.wikimedia.org/wiki/File:KF_25_Tee.jpg*).

Fig. 9.2. All-metal coupling (*Courtesy: Parker*).

high vacuum application. However, for UHV applications, stainless steel couplings with gaskets made of softer metals as shown in Fig. 9.2 can be used.

Most of the demountable joints in high vacuum systems with pipe diameters generally exceeding 25 mm are of flanged type with elastomer O-rings. Most commonly used elastomer in vacuum practice is neoprene with shore hardness 60±5 units. Working temperature limits of neoprene are 30°C to 140°C. For making O-rings for use on large diameter flanged ports, it is convenient to make them from neoprene cord of circular cross section. For this purpose, the cord of the appropriate length is cut and its ends are glued together by using rubber vulcanizing solution and cured. Other materials for gaskets include Viton A, Teflon, butyl rubber, Buna N. Permeation rates, gassing rates and range of temperatures for use are important

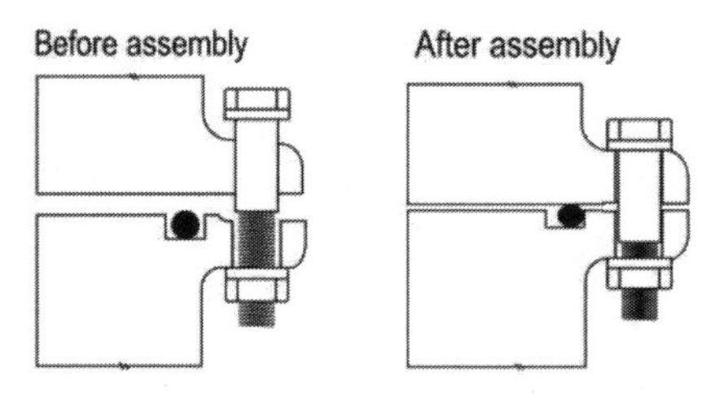

Fig. 9.3. Flanges with O-ring.

factors to be considered for selection of the gasket material. Useful information on these factors is given by Rosebury[2].

Figure 9.3 shows a commonly used flanged joint using the O-ring. Flanges with specially designed grooves for mounting O-rings are often used. It is necessary that a portion of the O-ring projects outside the groove prior to assembly. Grooves with the trapezium cross-section or 'dovetail' type of grooves are used on flanges located facing the ground to prevent the O-ring fall due to gravity. Some O-ring manufacturers provide dimensions of the grooves for use of the O-rings provided by them. Gaskets with specially designed shapes such as U-cups and Vee rings are also used.

For application in vacuum systems that are required to be exposed to high temperatures and UHV, gaskets made of relatively softer metals such as aluminum, OFHC copper and gold are used. Indium can be used up to about 150°C. OFHC copper and gold gaskets can be used in UHV systems that are required to be baked at 450°C. Metal gaskets[3] sandwiched between stainless steel flanges are shown in Fig. 9.4. The soft gasket material flows to make the seal as the flanges are tightened together thus facilitating vacuum-tight and high temperature resistant seals.

I used gold wire gaskets at Prof. Herb's laboratory at the University of Wisconsin during mid 1960's. A ring of gold wire of 0.5 mm cross-sectional diameter can be prepared by carefully fusing together the ends of piece of wire on a gas flame. This avoids the cutting and machining process involved in making OFHC copper gaskets.

9.2 Hoses and Bellows

High-pressure rubber tubes are used in fore-vacuum lines. These can bend and absorb vibrations to some extent. Flexible metal tubes with convoluted walls, called bellows are used for clean connections.

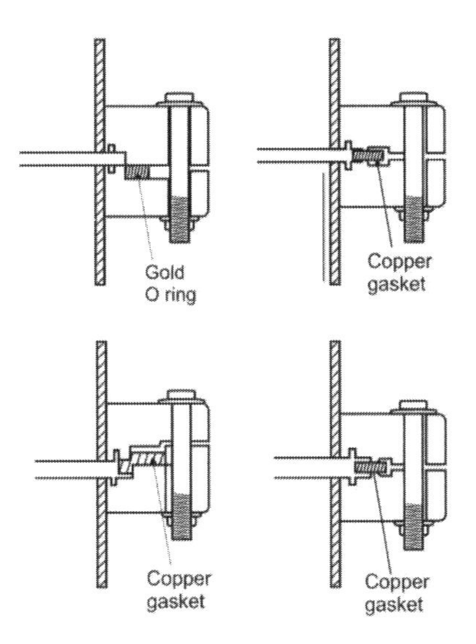

Fig. 9.4. Flanges for use with metal gaskets.

Bellows can also be used to communicate translational and rotary motions in vacuum. All-metal valves and mechanical manipulators employ stainless steel bellows. Thin flexible metal diaphragms can be employed to communicate small displacements in vacuum. Bellows can be single-ply or two-ply. Bellows are manufactured by using metal forming process or by edge-welding process. The edge welded bellows are more flexible and can undergo compression to higher extent. Bellows manufacturers provide data on compression, extension and spring force of the bellows.

9.3 Vacuum Valves and Traps

Vacuum valves are integral components of vacuum system as they provide necessary control for the flow of gas. The valves can be used for isolating different parts of the vacuum system. Different types of vacuum valves are selected depending upon the application. Valves can be manually operated, solenoid operated, electro-pneumatically operated or hydraulically operated. A needle valve, used for introducing controlled leak of gas has a plunger with a needle-like structure that engages into a hollow cavity and offers variable conductance depending upon the position of the needle in the cavity. A gate valve offers maximum possible conductance in its fully open position and offers complete isolation in the closed position. The gate valve consists of a plate with gasket and the plate can be lifted from or lowered to the valve seat to open or to close the valve and also to further slide across the port by a handle. The materials of the gaskets are selected depending upon the application. Isolation valves for UHV are bellows-operated, all-metal bakeable type and employ OFHC copper as nose pieces. Isolation in such valves is facilitated by biting of the soft copper nose piece by stainless steel knife edge. Alpert[4] and Lange[5] designed the bakeable valves for UHV. In such valves, it is important that the seal on the surface of the nose piece takes place at the same location every time the valves are closed. The nose pieces of the soft metal are required to be replaced after a certain number of close-open operations of the valves. Figure 9.5 shows the sub-assembly of an all-metal bakeable UHV valve developed at Prof. Herb's laboratory at University of Wisconsin.

Cold traps and baffle valves used with diffusion pumps to minimize contamination resulting from of oil vapors have been discussed in Section 6.1.6.

Cold traps are required to minimize the vapors of oil emanating from the pumps entering the vacuum system. A chevron baffle interposed on the top of diffusion pump serves effectively to minimize the back-streaming of oil vapors. The assembly of copper blades that is mounted in the baffle as illustrated in Fig. 9.6 is generally

Fig. 9.5. Sub-assembly of all-metal bellows-operated valve.

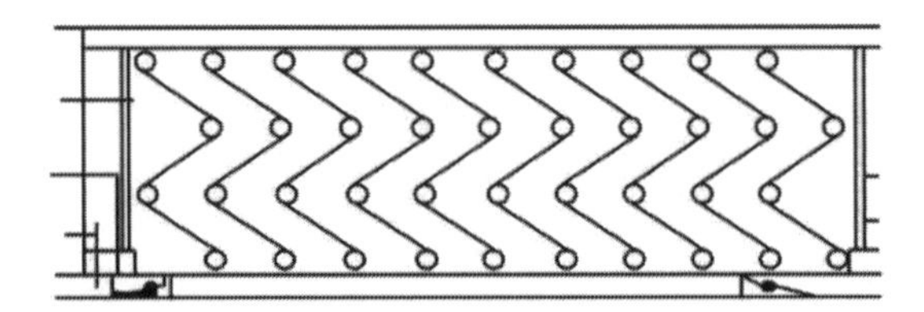

Fig. 9.6. Arrangement of blades in a chevron baffle.

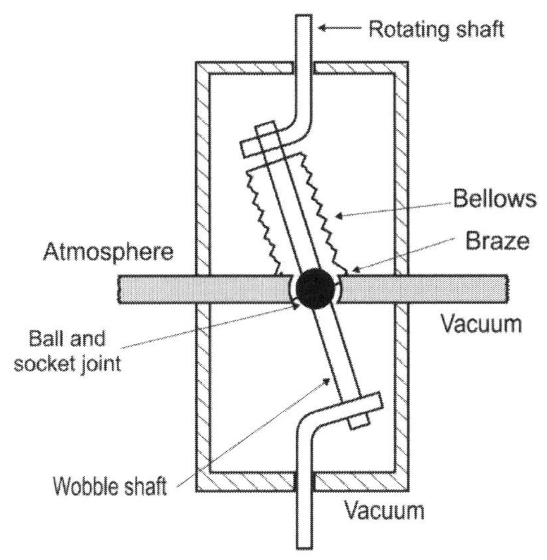

Fig. 9.7. Rotary motion feedthrough.

cooled by circulating water at about 15°C to facilitate trapping and condensation of oil vapors from the diffusion pump. Although the baffle causes reduction of the effective pumping speed, the advantage gained by minimizing the back streaming of oil vapors is significant.

9.4 Mechanical Manipulators for Vacuum

Figure 9.7 illustrates how rotary motion is communicated in vacuum using bellows.

All-metal bakeable mechanical manipulators that are capable of transmitting translational motion in vacuum in x, y and z directions and also capable of causing rotation and tilt of the study sample mounted in the vacuum chamber are used for some experimental surface science research works conducted in UHV. Manual and motorized versions of bellows operated manipulators with sample heating and cryo-cooling options are commercially available. Micrometers are used along with stainless steel bellows for precision positioning of the sample. Figure 9.8 shows a mechanical manipulator for UHV application.

9.5 Glass-to-Metal Seals

Glass-to-metal seals are used in metal vacuum systems primarily to connect glass components such as the glass gauge-heads to metal

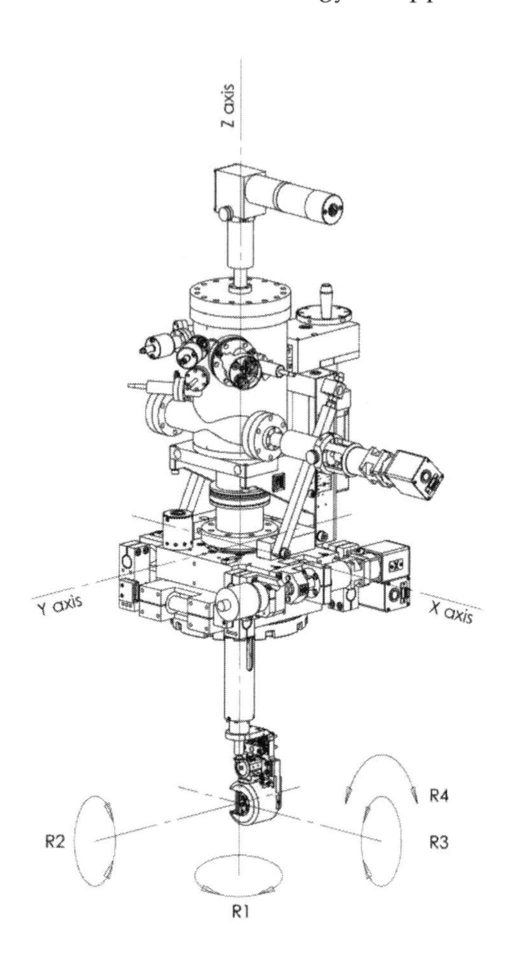

Fig. 9.8. Mechanical manipulator (*Courtesy: PREVAC*).

members of the system. Also, such seals are commonly used to feed electrical current or voltage into the metal system. The metals used in such seals are required to have their thermal properties such as coefficients of expansion, compatible to those of the glass. These include Kovar: Fe:Ni:Co (54:29:17); Fernico: Fe:Ni:Co:Mn (53.7:29:17:0.3); Rodar: Ni:Co:Mn:Fe (29:17:0.3:balance); Therlo: Cu:Mn:Al (85: 9.5:5.5).

Metals such as W, Mo, and Cu are also used for such seals. W is the most common metal for the 'press' or 'pinch' type of seals. Formation of one or more oxides upon heating facilitates the sealing process in case of the most glass-to-metal seals. The fused glass 'wets' these oxides and forms a glass–oxide metal transition. Borosilicate glass that is used for the most laboratory-ware including the ion gauges, offers high chemical stability, lower coefficients of thermal expansion, higher thermal shock resistance and an ideally high electrical resistivity.

For making a glass-to-metal seal, Kovar is polished, cleaned and dried. Furnace heat treatment is given to Kovar in wet hydrogen atmosphere at 900°C for one hour. Metal and glass parts are heated to about 850°C in air to obtain oxidized surface and the parts are brought together with pressure. Further, flame annealing of the seal is carried out. Generally, the Kovar portion turns light brown in good seals. Figure 9.9 illustrates a glass-to-metal feedthrough for electrical application in laboratory. During my work at BARC, Mumbai, it was a common practice to use a tubular 'graded' glass seal between Kovar tube and the borosilicate glass tube for sealing. The 'graded' glass comprises 'softer' glasses of different coefficients of expansion interposed between the 'hard' borosilicate glass tube and Kovar tube. Use of the graded glass assembly minimizes the strains that would be caused otherwise by directly making seal with borosilicate glass.

9.6 Ceramic-to-Metal Seals

Technical ceramics include oxides such as alumina, beryllia, ceria, zirconia; non-oxides such as carbide, boride, nitride, silicide and composites such as combinations of oxides and monoxides. High-alumina ceramic finds useful application in vacuum technology primarily due to its high electrical resistivity, high mechanical strength, high temperature resistance and non-porosity. The Al_2O_3 content is between 90% and 99% and the balance being SiO_2 to remove porosity. High-alumina (with alumina content of 95% and higher) insulators are commonly used in construction of vacuum electronic tube devices.

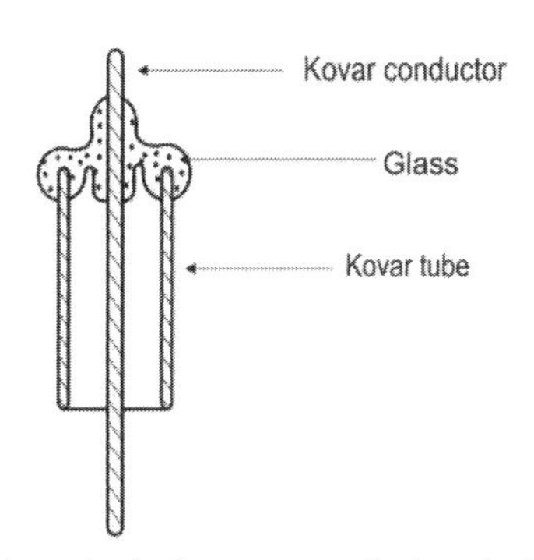

Fig. 9.9. A typical glass-to-metal electrical feedthrough.

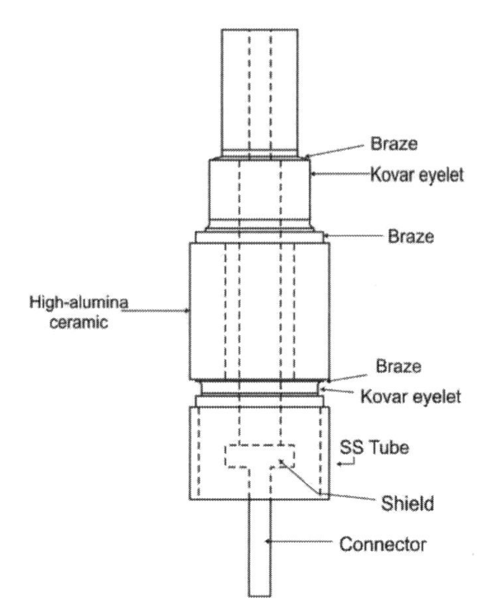

Fig. 9.10. Assembly of a ceramic-to-metal feedthrough.

Fig. 9.11. Flanged ceramic-to-metal feedthrough (Courtesy: CeramTec).

Ceramic-to-metal seals are employed in construction of UHV systems. Ceramic-to-metal feedthroughs are used for providing electrical connections in vacuum for high voltage, high current and RF applications. Figure 9.10 illustrates assembly of a ceramic-to-metal feedthrough. Figure 9.11 shows a typical ceramic-to-metal feedthrough welded to a flange.

Clarke et al[6] have published an excellent review on ceramic-to-metal joining. The 'Moly–Manganese' method[7] is a widely practiced technique of metal and ceramic bonding. The seal is achieved by metallizing of the ceramic, followed by brazing in vacuum or hydrogen. In this method, fine powders of Mo:80% and Mn:20% are thoroughly mixed together. The powder mix is then suspended in a volatile binder to prepare a paste of prescribed viscosity and thickness, which is then carefully applied to the region on the ceramic surface where bonding is desired. The coated assembly is then fired in hydrogen-atmosphere furnace at a temperature of about 1350°C (below the softening point of ceramic). The assembly is maintained at the maximum temperature for 20–30 min in the hydrogen atmosphere and then cooled to room temperature. Hydrogen firing facilitates sintering of the metal powders, combination with the ceramic and penetration into the surface thereby forming a strong bond. The thickness of the metallized coating is usually 0.025 mm to 0.038 mm.

A thin coating of nickel is further applied over the moly-manganese layer. The metal member and the coated ceramic are assembled together with brazing material OFHC Cu, Ag–Cu or Au–Cu in suitable form and brazed in hydrogen or in vacuum at a temperature depending upon the braze material. The entire process of sealing is required to be carried out in clean environments to prevent contamination that could hamper the process. The seals are later leak checked to ensure compatibility for vacuum application. A mechanical ‘pull test’ can also be conducted to ensure the required mechanical strength of the bonding.

As the metal has a higher coefficient of expansion than ceramic, cracks may develop in the ceramic during heating. The seal may open while cooling due to the stresses. It is therefore necessary to use an alloy that matches closely to the thermal expansion characteristics of ceramic. Kovar and Ni–Fe alloys are suitable choices for ceramic-to-metal seals. Figure 9.12 shows the thermal expansion characteristics of these alloys and the ceramic types.

The ‘active metal’ process requires that an ‘active’ element reacts with the ceramic to form a reaction layer between the ceramic and the molten braze material that will facilitate wetting of the ceramic. The active elements used for this process include Ti, Zr, Hf, V and Al. The best known of these active elements is Ti, which is used in many commercially available braze alloys. The active metal, in form of hydride, forms an oxide which is compatible with ceramic due to thermal action in an inert atmosphere. Kirchner[8] has described an active metal process using titanium hydride. The active metal process is a one-step operation. Brazing can be undertaken at a range of temperatures in controlled atmosphere, depending on the alloy used.

9.7 Fabrication Techniques

The vacuum system mounting involves utilization of demountable and permanent joints. Demountable joints including couplings and flanged

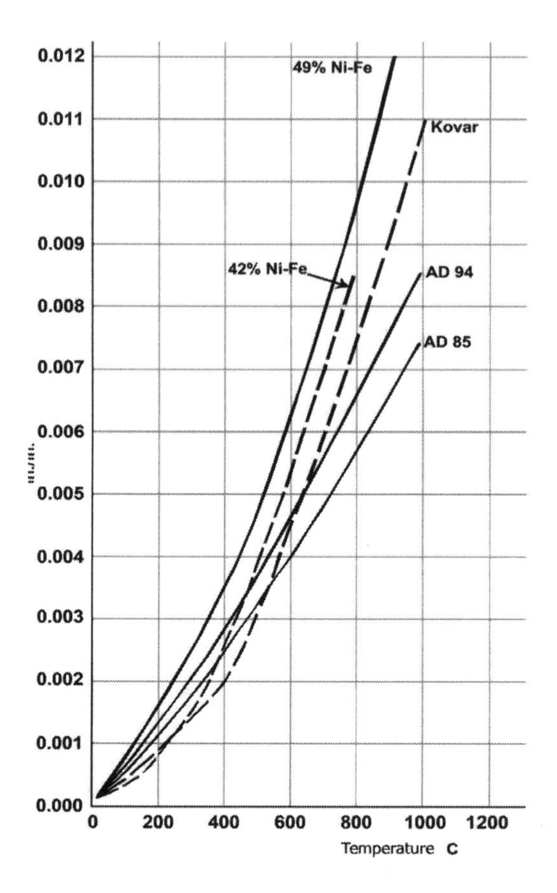

Fig. 9.12. Thermal expansion characteristics of Kovar and Ni–Fe alloys (*Courtesy: Coors Porcelain Company n/k/a CoorsTek, Inc.*).

joints have been discussed earlier in this chapter. Permanent joints are made by using soldering, brazing, welding and glass-blowing techniques. These also include glass-to-metal and ceramic-to-metal seals which have been discussed above.

9.7.1 Soldering and brazing

In soldering and brazing, similar or dissimilar materials are bonded together using a heating method and a filler metal without melting the base materials. The filler metal melts, wets the base materials, and flows by capillary action. Wetting of the base materials by the filler metals is achieved by using a suitable flux or by acoustic vibrations. The soldering process takes place below 450°C, while brazing takes place above 450°C but below the melting point of the base metal. The heating of the filler metal can be accomplished by various methods, including hot plate, induction, torch, and furnace. Use of soldering process is not recommended for vacuum applications, primarily due to the lead content that is involved in soldering materials. Joints made by using brazing have sufficient mechanical strength and generally can withstand bakeout temperatures up to 450°C. For vacuum applications, it is necessary to select appropriate brazing alloys with low vapour pressure. Thus metals such as Zn and Cd are excluded. Kohl[7] has given a list of brazing alloys suitable for vacuum applications.

It is essential that the metals to be joined are wetted by the braze material. This requires that the metals form alloy with at least one constituent of the braze material. Brazing is more often employed for small parts and is not suitable for large objects as it requires heating a broad surface to bring the filler material to its flow point, Unlike welding, lower temperatures are used in brazing. This facilitates joining of components without causing warpage or metal distortion. Brazing does not cause melting of one or both of the metals if the filler metal is metallurgically compatible with both base metals and has a melting point lower than either of the metals to be joined. It is essential to remove all oxides and surface oils before joining the metals as the contaminants interfere with the proper flow of filler metal and may reduce the joint strength. The surfaces to be joined need to be mounted securely to ensure an adequate capillary space between them for the flow of the molten brazing filler.

9.7.2 Vacuum brazing

Vacuum brazing is a materials joining technique that offers significant advantages including extremely clean, oxide-free, flux-free brazed joints of high integrity and strength. Temperature uniformity is maintained on the work piece when heating in vacuum. Vacuum brazing reduces residual stresses due to slow heating and cooling cycles thereby significantly improving the thermal and mechanical properties of the material. Brazing relies on the capillary action where the filler metal is drawn into the joint during the heating cycle. Although pure metals are sometimes used for brazing, most brazing materials are alloys of several different metals. A wide variety of brazing alloys are being developed, from those designed to increase the strength of joints and the corrosion resistance of materials, to those designed to achieve the same joint quality with less expensive materials. The braze material and its form/shape is selected depending upon the metals of the components/assemblies to be joined and their configurations. Vacuum braze materials are available in form of wire, washer, slug. Melting point and the flow point of the braze material are important parameters that govern the selection of the braze material. Some of the common braze materials include

- OFHC Cu (melting point: 1083°C)
- Au (melting point: 1063°C)
- Au–Cu eutectic alloy (Au: 80%, Cu: 20%, melting point/flow point: 889°C)
- Ag–Cu eutectic alloy (Ag: 72%, Cu: 28%, melting point/flow point: 780°C)

Vacuum brazing is employed for a number of applications in vacuum technology and in aerospace industry. The subject of vacuum brazing process and vacuum brazing furnace is discussed in chapter 12.2.

9.7.3 Welding

Welding involves joining of materials by coalescence. This is achieved by melting the objects with or without adding a filler material to form a pool of molten material which results into a strong joint upon cooling. In some cases, pressure is used along with heat, for welding purpose.

It is necessary to consider the weldability aspect of the materials to be welded. Welding process should not result into formation of undesirable alloys or constituents. Joints are required be free from discontinuities, porosity, shrinkage, or cracks. The welded joint should be able to withstand conditions such as extreme temperatures, corrosive environments, fatigue, and high pressures.

Fusion welding is the most common method of welding in which the base metal is melted. The filler metal may or may not be added. Heat is supplied by oxyacetylene gas, or electric arc or plasma arc or laser. During fusion welding, the molten metal is susceptible to oxidation. It is therefore necessary to protect the weld from the atmosphere by weld fluxes or inert gases or vacuum.

The commonly used fluxes include SiO_2, TiO_2, FeO, MgO and Al_2O_3. These fluxes form a gaseous shield to prevent contamination, serve as scavengers to reduce oxides, add alloying elements to the weld and influence the shape of the weld bead during solidification. Gases such as argon, helium, nitrogen, and carbon dioxide are used to provide a protective envelope around the weld area to avoid oxidation. Argon and helium are generally preferred.

Types of fusion welding include

- Oxyacetylene cutting/welding
- Shielded metal arc
- Metal inert gas (MIG)
- Tungsten inert gas (TIG)

Oxyacetylene welding is achieved by a flame formed by burning a mixture of acetylene (C_2H_2) and oxygen

In the case of shielded metal arc welding, an electric arc is generated between the tip of the coated electrode and the parent metal. The coated electrode produces a gas to control the atmosphere and provides filler metal for the weld bead. The tip of the electrode is melted by the arc. Small drops of metal enter the arc stream and are deposited on the parent metal. As molten metal is deposited, a slag forms over the bead which serves as an insulation against air contaminants during cooling. After cooling, the oxide layer is removed.

For inert gas welding, an inert atmosphere is maintained around the weld. The shielding inert gas such as argon or helium feeds through the welding gun, and shields the process from contaminants in the air. In case of tungsten inert gas welding, tungsten electrode serves as cathode. A plasma is formed between the tungsten cathode and the base metal. The plasma heats the base metal to its melting point.

Relative positions of the members to be welded are important to achieve satisfactory welding. In case of components for vacuum application, it is essential not to expose to vacuum any trapped volume that would result from welding as it can serve as a source of 'virtual leak' resulting from gradual release of the trapped gas over a long period. Examples of proper positioning of components for welding are shown in Figure 9.13.

9.7.4 Electron Beam Welding

Welding in vacuum offers high-quality welds. Electron beam welding (EBW) is employed in nuclear applications involving metals such as Zr, Hf and Ti. Such welding minimizes impurities by a significant factor. In this method, a high-velocity beam of electrons in vacuum is focused on the materials to be joined. The objects melt and flow together as the kinetic energy of the incident electrons is transformed

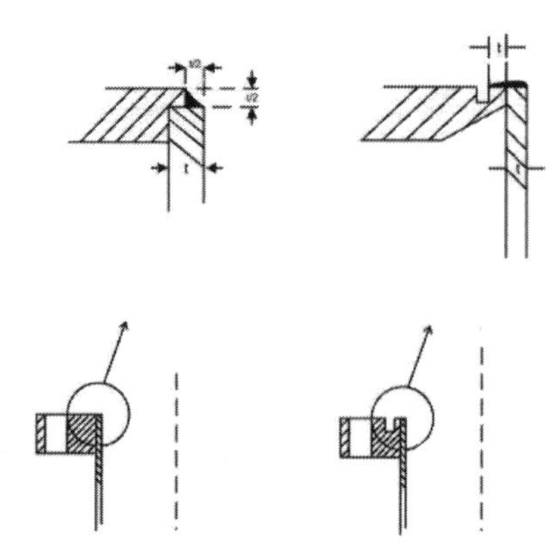

Fig. 9.13. Positioning of components for welding.

into heat. EBW is generally performed under vacuum conditions to prevent scattering of the electron beam.

The EBW system is composed of an electron beam gun, a power supply, control system, motion equipment and vacuum welding chamber. The electron beam energy can vary between 30–200 keV with electron current variation between 0.1 mA to 1 A depending upon the job. 60 keV beams are commonly used for manufacture of small components including battery cans, medical components, aneroid capsules, latches, bellows, relays and electronic parts including transducers. Applications in the automotive industry include airbag inflators, gearbox components, transmission parts and turbo-charger shaft wheel assemblies. 150 keV beams are used extensively in the aero engine, aerospace and power generation industries. The electron beam stream and workpiece are manipulated by means of precise, computer driven controls, within a vacuum welding chamber, thus eliminating oxidation, contamination.

The characteristics of electron beam welds include

- High depth-to-width ratio
- Inert atmosphere (vacuum)
- Near parent metal strength
- Minimum component distortion
- High accuracy

9.8 Cleaning

Cleaning of the vacuum components prior to their assembly in the vacuum system and surface processing treatments in vacuum with appropriate methods are essential for satisfactory performance of the processes to be carried out in vacuum. Proper cleaning eliminates the presence of contaminants that can hamper the processes in vacuum. Even a fingerprint on the surface of a component could give rise to undesirable outgassing when the component is exposed to high vacuum. Surface treatments and cleaning methods commonly used in vacuum practice are discussed here.

Dust, lint, soot from air and human body are generally the common environmental contaminants. Mineral salts, oils, greases, fatty acids, oxides, silicates, sulfides, nitrides, borides and gases are the other major contaminants.

The U.S. Environment Protection Agency (EPA) has published[9] useful information on cleaning and degreasing processes.

9.8.1 Mechanical Cleaning

In the abrasive blasting process, a forcibly propelled stream of abrasive material such as sand or fine glass beads, is directed against a surface under high pressure to smooth a rough surface (or to roughen a smooth surface), or to remove surface contaminants.

In the tumble cleaning process, metal tumbling is used to burnish, deburr, clean, remove rust, polish, brighten, and to prepare parts for further finishing. The process involves rotation of a horizontal barrel filled with the components to be processed. Upon rotation, the components rise until the uppermost layer slides down to the other side due to gravity. The barrel can also be fitted with vanes, typically made of rubber, that run along the inside of the barrel. The vanes catch and lift the parts, which eventually slide down or fall as the barrel turns.

Wet processes employ a compound, lubricant, or barreling soap that is added to facilitate the finishing process, prevent rusting, and to clean components. Common media materials include sand, granite chips, slag, steel, ceramics, and synthetics.

9.8.2 Chemical Cleaning

Vapour degreasing involves cleaning of parts by condensing solvent vapors on components and assemblies. It employs hot vapors of chlorinated or fluorinated solvent to remove organic contaminants. A vapour degreaser comprises a stainless steel tank with a heated solvent reservoir at the bottom and a cooling region near the top. Heaters located at the bottom of the tank cause the solvent medium such as trichloroethylene to boil, thereby generating hot solvent vapour. The hot vapour displaces the air and fills the tank up to the cooling region at the top of the equipment. The hot vapour is condensed at the cooling region. The temperature difference between the hot vapour and the cool components facilitates the vapour to condense on the surfaces of the components and assemblies suspended in hanging stainless steel baskets and to remove the contaminants by dissolving.

Acid or caustic etching, chemical polishing processes are also used in chemical cleaning. In the boil degreasing method, usually hydrocarbon or a chlorinated hydrocarbon, or alkaline aqueous cleaner is heated up to boiling point in a bath and the parts are suspended in the boiling liquid. Fats and waxes are thus removed from the parts. The continuous movement of the bubbles causes a dispersing of the dirt. In many cases, this method is used as a preceding step in a vapour degreaser.

9.8.3 Electrochemical Cleaning

Electropolishing and passivation processes are used in electrochemical cleaning. Electropolishing is used for cleaning and rendering bright finish to metal components. This process removes the surface of the metal and is reverse of electroplating. Generally, stainless steel components used in UHV and XHV systems are electro-polished prior to their assembly. In the passivation process, the outer-layer of base material is shielded in form of microcoating, or oxidation. Passivation involves use of a light coat of a protective material, such as metal oxide, to offer resistance against corrosion.

9.8.4 Ultrasonic Cleaning

This process involves high frequency (20–400 kHz) agitation of the components and assemblies immersed in the cleaning medium. An ultrasound generating transducer built into the chamber, generates ultrasonic waves in the fluid. This creates compression waves in the liquid of the tank. The agitation is produced by the rapid creation and collapse of several small vapour bubbles that results in removal of adhesive layers of contaminants on the surfaces of the components and assemblies. This method is also useful for dislodging minute particulate contaminants that remain trapped in the narrow, deep crevices on the surfaces. The fluid medium can be trichloroethylene and other cleaning chemicals.

9.8.5 Thermal Cleaning Processes

Vacuum bakeout, hydrogen and vacuum firing, processing with air, nitrogen or oxygen are the processes employed in thermal cleaning. Section 11.4 discusses these processes.

9.8.6 Commonly Used Cleaning Chemicals

Trichloroethylene (C_2HCl_3)

Trichloroethylene is a chlorinated hydrocarbon commonly used as a solvent in ordinary and vapour degreasing techniques. Some of its properties are: boiling point: 86.7°C; distillation range: 86–87.5°C; flammability: nonflammable.

Acetone (C_3H_6O)

Acetone is colorless, mobile. It is also used as a cleaning chemical. However, it is not recommended in vacuum practice due to its health hazards and inflammable property. Its boiling point is 56°C.

Ethyl alcohol (C_2H_6O)

Also called ethanol, it is a volatile, flammable, colourless liquid and is used in vacuum practice, particularly to remove traces of water used during the final stages of cleaning process. It drives away water traces during its evaporation; boiling point: 78.37°C.

Hydrofluoric Acid (HF)

Hydrofluoric acid is a mineral acid used in dilute solutions to clean aluminum, glass, molybdenum, stainless steel and titanium. It should be used with caution in a ventilated hood to prevent injuries.

Methods of cleaning the commonly used metals in vacuum practice are described below.

Cleaning of copper

In the bright-dip process, copper components are immersed in the bath as given below.

Water: 1960 ml
H_2SO_4: 1730 ml (sp. gr. 1.84)
HNO_3: 285 ml (conc.)
HCl: 10 ml (conc.)

Cleaning of stainless steel

Electropolishing is generally used for giving a bright finish to stainless steel components. One of the baths uses the following chemicals:
Phosphoric acid: 5 parts by volume
Sulphuric acid: 4 parts by volume
Glycerin (USP): 1 part by volume
Current density: 0.5A cm^{-2}
Voltage: 9 V

Cleaning of Aluminium

- Degrease, if oily
- Use 15–60 $grams.liter^{-1}$ NaOH or Na_2CO_3
- Bright-dip for a few seconds at room temperature in a bath containing:

H_2SO_4: 7.5 liter
HNO_3: 3.8 liter
Water: 1 liter
HCl or (NaCl): 14 gram

Cleaning of glass

HF: 4 vol (conc.)
HNO_3: 33 vol (conc.)
Water: 60 vol
Detergent without filler: 2 vol
At room temperature

Cleaning of Kovar and nickel

Acetic acid, glacial: 750 ml
HNO_3: 250 ml (conc.)
HCl: 3 ml (conc.)
Temperature: 60°C

Cleaning of Tungsten

- Immerse in freshly prepared solution of sodium hypochloride 15%..1 vol and water..5 vol
- Boil 5 to 8 minutes
- Rinse in running tap water and in deionized water
- Rinse in reagent grade methanol

Cleaning of molybdenum

Immerse in a solution of

- Tap water..6 vol
- HNO_3 (conc.)...13 vol
- HF (conc.)..1 vol

Immerse in a solution of

- Deionized water...65 ml
- CrO_3...20 gm
- H_2SO_4 (sp. gr. 1.84): ...35 ml

(Acid to be added slowly while stirring)
Drain and rinse in hot running tap water
Boil for 5 minute in a solution of

- Tap water..6 vol
- NH_4OH (conc.)..2 vol

Drain off and rinse in running tap water
Rinse in deionized water
Boil for 10 minutes in fresh deionized water
Rinse in fresh methanol (reagent grade)
Drain and dry in oven at temperature <80°C.

Cleaning of titanium

Pickle in a solution of 20–30% HNO_3 with 2% HF at 54°C.

9.8.7 Clean Rooms

In order to ensure contamination-free, clean surfaces, handling and mounting work for components and assemblies of certain electronic components, critical vacuum devices including vacuum interrupters is required to be carried out in dust-free, contamination-free, temperature and humidity controlled 'clean rooms'. The walls and the flooring of the clean rooms demand the use of special materials. The tables, trollies and fixtures to be used in the clean room are generally of stainless steel. Clean rooms are classified by their capability of handling specified maximum dust content per unit volume of the room. The dust content is controlled by a series of blowers, HEPA (High Efficiency Particulate Air) and other types of air filters. The ambient air in a typical urban environment contains 35 000 000 particles per cubic meter in the size range 0.5 μm and larger in diameter. An ISO 1 clean room allows only 12 particles per cubic meter of 0.3 μm and smaller. The dust content can be monitored by special measuring equipment based on scattering of light by dust particles. The relative humidity and the temperature is controlled in clean rooms. Entry of personnel into the clean room area is restricted only to the personnel working in that area and the personnel are required to wear clean, lint-free, suits with hoods, gloves and slippers that cover the entire body and hair. The personnel are required to enter the clean area after exposure to air shower room to remove the dust on the suit.

References

1. Hilton, M. R. and P. D. Fleischauer, "Lubricants for High-Vacuum Applications", Space and Missile Systems Center, Air Force Material Command, Engineering and Technology Group, (1993).
2. F. Rosebury, "Handbook of Electron Tube and Vacuum Techniques", Addison Wesley Publishing Company, Inc. USA (1965).
3. W. J. Lange and D. Alpert, Rev. Sci. Instrum. **28**, 726 (1957).
4. D. Alpert, Rev. Sci. Instr. **22**, 536 (1951).
5. W. J. Lange, Rev. Sci. Instr. **30**, 602 (1959).
6. J.F. Clarke, J. W. Ritz, and E. H. Girard, Technical Report, Air Force Materials Laboratory (AFML), TR-65-143 (1965).
7. W. H. Kohl, "Materials and Techniques for Electron Tubes", Reinhold Publishing Corporation, New York (1960).
8. K. Kirchner, 1963 Vac. Symp. Trans. 170 (1964).
9. "Guide to Cleaner Technologies", US EPA, Office of Research and Development, Washington DC 20460, EPA/625/H-93/017, February (1994).

10

Leak Detection

Atmospheric air rushes into the vacuum system wherever it finds an access to enter as a result of the pressure difference. Such leaks are primarily due to the demountable and permanent joints not properly sealed. Leaks through gaskets or through brazed/welded joints are common in vacuum practice. Leak detection involves finding location of the leak and determining magnitude of the leak. This helps elimination of the leak by taking corrective actions in an effective manner so that the desired degree of vacuum can be achieved.

Leak detection can be achieved by two methods. In one method, the object is filled with compressed gas, sealed and then the location from where the compressed gas leaks out is detected. In the other method, the object is evacuated, then sprayed externally with a probe gas or fluid. A suitable sensor such as a vacuum gauge or mass spectrometer is used in the vacuum system that can give an indication, location and magnitude of the leak.

10.1 Methods Using Pressurization

In the pressure test, the object is temporarily blanked by using a bolted flange with a gasket and is further pressurized to a pressure marginally higher than the atmospheric pressure using a compressor. The component is then brushed with a soap solution over its surfaces. The region of soap bubble formation over the surface gives the location of the leak. The pressure test can be used for locating coarse leaks due to faulty brazed or welded joints. Alternately, the pressurized object is immersed in water and is observed for the escaping bubbles from the leaky regions. The former method has a higher sensitivity than the water immersion method. It allows detection sensitivity up to about 10^{-7} Pa · m^3 · s^{-1}. These methods are crude and suitable particularly for small volume objects.

In another method, the object is filled with a high pressure gas, isolated from the gas supply and the internal pressure P is monitored over time. The pressure drop is then indicative of leak. If ΔP is the fall in pressure in time t, then the leak rate is given by $\Delta P \times V \times t^{-1}$ where V is the volume of the system containing the object.

In the sniffer technique a detector probe is used to sense and measure leaks from an object previously filled and pressurized with a tracer gas. The gas escaping from the leak is sucked by the sniffer. The sniffer accommodates a device of adjustable leak rate. The probe is moved over the object to detect the leak. Accuracy of leak detection is determined by the speed, distance from the object and the probe sensitivity. Depending on the tracer gas, sniffer can locate a leak on an object with a sensitivity of about 10^{-8} Pa · m^3 · s^{-1}. If a containment hood can be made to surround the object pressurized with the tracer gas, the leaked tracer gas accumulates in the hood and is sensed by the detector. The halide diode can be used as a detector for gases with halogens.

In another method, a mass spectrometer tuned to helium is maintained at high vacuum (up to 1×10^{-2} Pa). One arm of the sniffer is connected to vacuum while the detection probe, fitted with a calibrated orifice, is exposed to the atmosphere. Air, along with helium enter the mass spectrometer through the orifice. The helium concentration in the mass spectrometer is monitored and the leak rate is determined. In this method, the object to be tested is pressurized with helium and the sniffer is moved around the suspected leak sites on the object. Helium, due to its low mass, has high conductance and leaks out much faster. Also, helium being an inert gas, does not cause any chemical reaction. The partial pressure of helium in the atmosphere is too small, about 5×10^{-1} Pa and has a negligible background.

10.2 Methods using Evacuation

In one of the methods, the system is evacuated to low pressure, then isolated from the pump and the rate of increase in pressure is monitored. Figure 10.1 shows the rise in pressure with time in such case. Though the theoretical sensitivity of this method is about 1×10^{-3} Pa · s^{-1}, the surface out-gassing can be high enough to cause

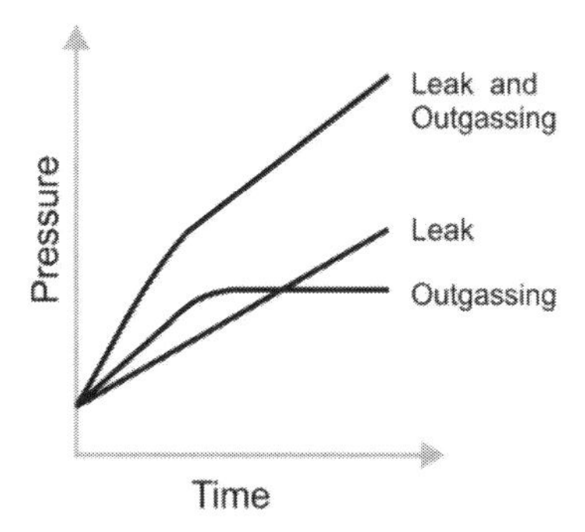

Fig. 10.1. Rise of pressure in chamber after isolating the vacuum pump.

the pressure to increase in a manner similar to that in case of a leak. The variation of the rate of rise of pressure with time can give an indication of presence of leak and/or outgassing. This technique also can detect the total system leak and not its leak location.

In a halogen leak detector, an evacuated vessel is connected to a halogen detecting instrument and is sprayed with halogenated gas such as CFC, HCFC and HFC. Its sensitivity is about 5×10^{-8} Pa · m^3 · s^{-1}.

At Prof. Herb' laboratory at the University of Wisconsin, I use used helium as the probe gas for locating leaks in an all-metal UHV system comprising an orbitron pump and a B-A gauge. Upon spraying helium through a nozzle/syringe around the suspected leaky joints such as flanges, welds, brazes, the helium pressure, would build up in the event of leak in the UHV system due to much lower pumping speed of helium offered by the orbitron pump, thus indicating sudden rise in the pressure reading by the gauge.

The mass spectrometer using helium as the probe gas is the most commonly used method for detection of leaks in high vacuum and UHV systems. In this method, the component or the assembly to be leak checked is connected to the leak detector equipment that includes a vacuum system with a suitable type of mass spectrometer. The component or the assembly to be leak checked is evacuated by the vacuum system of the equipment. Helium gas oozing out from a small syringe shaped nozzle is then sprayed over the possible locations of the suspected leaks. In the event of leak, the mass spectrometer in the equipment registers the presence of helium to which it is tuned. The smallest leak rate or the minimum leak that can be detected by the leak detector is termed as the sensitivity. The maximum sensitivity of the equipment is about 1×10^{-11} Pa · m^3 · s^{-1}. Such leak detectors are commonly used in research laboratories

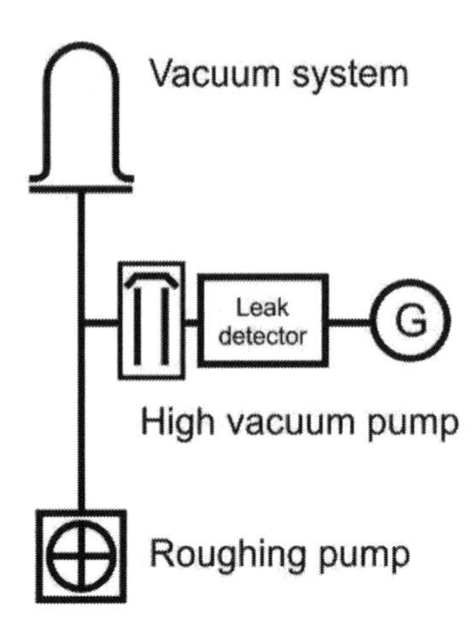

Fig. 10.2. Schematic diagram of a helium counter flow leak detector.

and for leak checking small size components and assemblies in production lines. The leak detector equipment can be calibrated by using standard calibrated leaks using high-purity helium. These are commercially available in different flow rate ranges. Leak elements are generally in form of a micro-tube capillary. The hood test is used to check an overall leak. In this method, the test object is covered with a plastic hood filled with helium so the exterior of the test object is exposed to helium which enters through the locations of the leak on the evacuated test object and is detected.

Becker[1] developed a method of helium counter-flow for leak detection in which helium, upon spraying on the pre-evacuated object under examination, flows counter to the pumping direction through the turbo-pump (or diffusion pump) to the mass spectrometer mounted near its inlet port, where it is detected. Figure 10.2 illustrates the vacuum system for this method. The sprayed helium, after passing through the leak in the object, finds an easier access to the inlet port of the turbo-pump/diffusion pimp through its exhaust port due its smaller size and lower compression ratio. The air that also gets sprayed along with helium, cannot counter-flow into the inlet port of the pump due to its higher compression ratio while helium arrives there at relatively higher partial pressure.

The mass spectrometer enables helium to be detected in the object even at pressures of less than 100 Pa. A leak rate between 10^{-2} and 10^{-12} Pa · m^3 · s^{-1} can be covered by this method.

10.3 Leak Detector Calibration

The leak detector is required to be calibrated to determine leakage rates. This can be achieved by using a commercial test leak, a vessel filled with helium having a shut-off valve that releases the gas at a known rate. Generally, this is built in the assembly of the leak detector or can be coupled with the intake side. The calibration is achieved by using a suitable working cycle. It is advisable to calibrate the device every time before taking the measurement.

Reference

1. W. Becker, Vak. Tech. **8**, 203 (1968).

11

Extreme High Vacuum

11.1 Factors Governing Ultrahigh and Extreme High Vacuum

Ultrahigh vacuum (UHV) corresponds to the pressure range between 10^{-10} Pa and 10^{-7} Pa and extreme high vacuum (XHV) corresponds to the pressure range below 10^{-10} Pa. At present, the application of XHV is limited to few applications such as certain particle accelerators, storage rings, space simulation, processing of semiconductor devices, and a few other specialized applications. Unlike high vacuum, UHV and XHV practices demand much more careful selection of materials, components, pumps, gauges and processing techniques. While obtaining UHV is common practice now, achieving XHV demands utilizing certain additional processing techniques to further minimize the outgassing rates from the surfaces exposed to vacuum. The pumps utilized for production of XHV include the positive displacement type such as TMP, diffusion pumps and capture pumps such as cryo-pumps, cryo-sorption pumps, getter, and getter–ion pumps. To measure pressures in the XHV range, it is essential to use ionization gauges that offer very low X-ray induced residual currents and minimum electron-stimulated desorption of ions and neutrals.

Assuming that the maximum pumping speed that can be applied to a vacuum system of arbitrary size is approximately proportional to the surface area, Readhead[1] has shown that the ultimate pressure P_u is independent of the system volume and the surface area. He explained that the total outgassing rate Q is given by

$$Q = Q_T + Q_d + Q_h + Q_{st}$$

where Q_T is the total thermal outgassing rate = $q_T \times A \times f$ (q_T is the outgassing rate per unit area, A is the surface area, f is the surface roughness factor); Q_d is the diffusion rate through envelope = $q_d \times A \times f$ (q_d is the diffusion rate per unit area); Q_h is the outgassing rate through heated components; Q_{st} is the desorption rate due to electron, ion or photon stimulation

He presumed that in general,

$$(Q_T + Q_d) \gg Q_h + Q_{st}$$

Equation (2.34) gives the basic pumping equation as

$$V\frac{dP}{dt} = -PS + Q$$

where V is the volume to be pumped; P is the pressure at time t; S is the pumping speed; Q is the total rate of influx (throughput) of gas.

Also, at ultimate pressure P_u, $\frac{dP}{dt} = 0$ and

$$P_u = \frac{Q}{S} = \frac{\left[(q_T + q_d)Af + Q_h + Q_{st}\right]}{S}$$

He further explains that in general, the maximum pumping speed that can be applied to a vacuum system of arbitrary size is approximately proportional to the surface area ($S_{max} = kA$). Thus the ultimate pressure P_u with maximum pumping speed is approximately

$$P_u = \text{const} \times \left[(q_T + q_d)f + (Q_h + Q_{st})/A\right] \quad (11.1)$$

$$\text{If} \quad Q_T + Q_d \gg Q_h + Q_{st}$$

The ultimate pressure with the maximum pumping speed is independent of the system volume and the surface area

$$P_u = \text{const} \times (q_T + q_d)f \quad (11.2)$$

These equations are valid if the gas density is uniform throughout the vacuum chamber. In practice, most extended XHV systems tend to exhibit non-uniform pressure distributions. Extremely small leak rates need to be measured for XHV using helium leak detectors of sensitivity better than 10^{-12} Pa · m^3 · s^{-1}. This can be achieved by allowing the helium to accumulate in presence of NEG for pumping of active gases. The leak rate can be determined by the rate of rise of the helium ion current[2] in the mass spectrometer. The sensitivity of this method is 10^{-16} Pa · m^3 · s^{-1}.

11.2 Pumps for XHV

Various pumping combinations have been used for producing XHV depending upon the application.

Santeler[3] achieved pressures of the order of 10^{-12} Pa by using a diffusion-getter pump combination. While using the diffusion pump, the fore-pressure is required to be maintained very low to avoid hydrogen in the fore-vacuum back-diffusing through the vapour jets, thereby lowering the effective hydrogen pumping speed. Also, the back-streaming of oil from the diffusion pump and the rotary pumps requires to be prevented from entering the XHV region by employing careful trapping. Enosawa et al[4] attained a pressure of 10^{-9} Pa by using an assembly of two TMPs with magnetically suspended rotor on a common shaft to provide hydrogen compression ratio of 5×10^8 and maximum backing pressure of 50 Pa. It was observed that the ultimate pressure can be further lowered to $<10^{-10}$ Pa by reducing the rotor temperature[5].

The specially designed TMP with the magnetically suspended rotor and backed by the diaphragm pump that can provide a high backing pressure can be used for XHV application, particularly for relatively small systems. Abundance of hydrogen at XHV demands enhancement of the low compression ratio of hydrogen of about 10^4 in the TMP.

Cho et al[6] achieved pressures in the low 10^{-10} Pa range as measured by an extractor gauge in a stainless steel test chamber of 6800 cm^2 surface area with a turbo-molecular pumping system. Low backing pressures were obtained by a molecular drag pump with the membrane pump. Benvenuti[7] achieved pressure of 5×10^{-12} Pa in a 3 m long section of an accelerator using Zr–V–Fe NEG strip, a sputter ion pump, and a titanium sublimation pump. Evaporable and non-evaporable getter pumps are commonly used in XHV applications. As these pumps cannot pump rare gases and methane, other pumps such as sputter-ion pumps and cryo-pumps are required to be used in combination with the getter pumps. The high pumping speed for hydrogen provided by (NEG) is advantageous as the predominant gas at XHV is hydrogen. NEGs located in close proximity with beam lines in particle accelerators and storage rings are particularly effective in maintaining low pressure. Benvenuti[8, 9] sputter coated the interior of a stainless steel system with a thin (~1 μm) film of getter material. Ti, Zr, V and their binary alloys have been used as getter materials. The coated surface is transformed from a gas source to a pump by an in-situ bakeout at temperatures of 250–300°C. Pressure of 10^{-11} Pa has been produced by using this method.

Cryo-condensation and cryo-sorption types of pumps also find application in XHV. Historically, there are instances when XHV pressures were possibly attained[10,11] in relatively small systems in experimental research work using cryo-condensation but could not be measured due to lack of measurement capabilities. Tsukui et al[12] evacuated the main analysis chamber by a TSP and a cryo-sorption pump at 4.2 K to obtain the XHV condition that could be maintained for longer than 200 h. The beam pipe in the centre of the superconducting magnet coils of the Large Hadron Collider at CERN is maintained in direct contact with the helium bath at 1.9 K (cold-bore beam pipe)[13]. At this temperature, the vacuum chamber wall becomes a very efficient cryo-pump.

11.3 Gauges for XHV

Hot-cathode ionization gauges are used for measurement of pressure in the XHV region. It is essential that the X-ray limit of the gauge is sufficiently low and that the gauge is capable of separating the gas-phase ions from the electron-stimulated desorbed (ESD) ions. Such gauges include

- Lafferty gauge[14,15]
- Extractor gauge[16]
- 90° Bent-beam gauge (Helmer gauge)[17]
- 180° Bent-beam gauge[18]
- 256.4° Bent-beam gauge [19]
- Bessel box (A-T gauge)[20]

11.4 Reduction of Outgassing

For attaining XHV pressures, it is essential to minimize outgassing rates of the materials used in the construction of vacuum systems. The residual gas at XHV in metal systems is predominantly hydrogen. It is important to reduce the outgassing rate of hydrogen to achieve pressure in the XHV range. The outgassing rates can be reduced by a number of methods.

High-temperature bakeout

A high-temperature (up to 450°C) bake of the vacuum system can be used by employing an oven that can supply heat with uniform temperature distribution. It has been discussed previously in Section 3.1.4 why high-temperature bakeout is essential to minimize the outgassing rates from the walls of the system exposed to vacuum. The maximum bakeout temperature is limited by the materials used in the system. Bakeout at 450°C for a few hours is adequate to significantly reduce the outgassing rates from the walls of all-metal systems with glass gauge-heads. Bakeout temperatures of higher than 450°C could be harmful for borosilicate glass generally used for gauge-heads. Uniform heating of the system has to be ensured to avoid temperature gradients. In metal and glass systems, mostly water vapour, CO and H_2 are released during the bakeout. Outgassing rates after bakeout can be estimated upon isolating the pump from the rest of the system and then by measuring the rate of rise of pressure in the system of known volume and surface area. Hobson[21] obtained outgassing rate of 2.5×10^{-13} Pa·m·s^{-1} for alumina silicate glass after baking at 500°C for 18 hours, followed by 600°C for 2 hours, 700°C for 2 hours and 500°C for 10 hours.

High-temperature firing in air

The oxidation of stainless steel, believed to form an oxide barrier to hydrogen diffusion from the bulk, has been shown by many experimenters to reduce the outgassing rate to the 10^{-11} Pa·m·s^{-1} range. Recent measurements by Bernadini[22] et al have shown that firing stainless steel in air at about 400°C drives out most of the hydrogen in the bulk but the presence of the oxide layer does not reduce outgassing. A vacuum re-melted type 316 L stainless steel was oxidized in air at 100°C and 250°C by Ishikawa and Odaka[23]. Subsequent baking produced a surface of outgassing rate of 4×10^{-11} Pa·m·s^{-1}.

Vacuum firing

Grosse and Messer[24] observed outgassing rate of 1.6×10^{-13} Pa·m·s^{-1} (H_2) for stainless steel after glass bead blasting, followed by vacuum firing at 550°C for 3 days, further followed by bakeout at 250°C for 24 hours. They also observed the outgassing rate of 6×10^{-14} Pa·m·s^{-1} (H_2) for OFHC Cu after vacuum firing at 550°C for 3 days, followed by 250°C bakeout for 24 hours. Marin et al[25] found that the outgassing rate after vacuum firing was 3×10^{-11} Pa·m·s^{-1} whereas the rate after air firing the same type of sample was 1×10^{-12} Pa·m·s^{-1}. Fremery[26] has measured an outgassing rate of 1×10^{-12} Pa·m·s^{-1} after vacuum firing at 960°C.

TiN deposition

Titanium nitride has been deposited on electrolytically polished stainless steel using a hollow cathode discharge[27] after vacuum firing twice at 430°C for 100 h and 500°C for 100 h. The outgassing rate at room temperature was lowered by a factor 10 by the TiN layer to 1×10^{-13} Pa·m·s^{-1} (H_2). 100 nm thick film of of titanium nitride (TiN) was coated on walls of the stainless steel vacuum chambers[28] of the 248 m accumulator ring of the Spallation Neutron Source (SNS) with an objective of reducing the secondary electron yield. The outgassing rates of several SNS half-cell chambers were measured with and without TiN coating, and before and after in-situ bake. The coating resulted in reducing, the in-situ baked H_2 outgassing rate by about 25%, indicating an evidence of a permeation barrier to hydrogen diffusion.

Machining in an inert atmosphere

Ishimaru[29] has described an aluminium alloy system where the chamber was machined in an atmosphere of dry oxygen and argon after bakeout at 150°C for 24 hours. The outgassing rate was about 10^{-10} Pa·m·s^{-1}, and ultimate pressure of 4×10^{-11} Pa was claimed.

Ti–steel alloy

Kurisu et al [30] observed outgassing rates lower than 6×10^{-13} Pa·m·s^{-1} for Ti–steel alloy. They explain that in the Ti alloy a natural TiO_2 thin layer on the surface acts as an enhanced diffusion channel for dissolved gases, mainly hydrogen, thus giving a very low outgassing rate.

Be–Cu alloy

Watanabe[31] attained low outgassing rate of 5.6×10^{-14} Pa·m·s^{-1} (H_2) for a Be–Cu alloy. After a pre-bakeout treatment in UHV at 400°C for 72 hours, the surface of 0.2% BeCu is covered by a BeO layer.

This layer serves as a barrier to the processes of oxidization and permeation of hydrogen. Further, using an in-situ bakeout treatment, the temperature is ramped up to 150°C and immediately ramped back down.

NEG coating

A coating of Ti/Zr/Va NEG [32] activated through a bakeout temperature of ~200°C has been used to achieve a low outgassing rate.

Plasma glow discharge cleaning

Plasma or glow discharge cleaning is utilized to remove solid surface contaminants with energetic plasma or glow discharge in vacuum. Species, generally, argon and oxygen, or air, hydrogen/nitrogen mixture are used. Interaction of plasma with surface causes breaking of bonds of organic contaminants of the surface and the removal of the fragments by vacuum. Intense argon glow discharge cleaning (10^{17} A^+ ions · cm^{-2}) was required to obtain a low desorption yield for a stainless steel surface, releasing an equivalent of 77 monolayers of CO.[33]

References

1. P. A. Redhead, "Extreme High Vacuum", Proc. the CERN Acceleration School, Snekersten, Denmark, CERN Report, 213, edited by S. Turner, (1999).
2. L. E. Bergquist and Y. T. Sasaki, J. Vac. Sci. Technol. A **10**, 2650 (1992).
3. D. J. Santeler, J. Vac. Sci. Technol. **8**, 299 (1971).
4. H. Enosawa, C. Urano, T. Kawashima, and M. Yamamoto, J. Vac. Sci. Technol. A **8**, 2768 (1990).
5. H. Ishimaru and H. Hisamatsu, J. Vac. Sci. Technol. A **12**, 1695 (1994).
6. B. Cho, S. Lee, and S. Chung, J. Vac. Sci. Technol. A **13**, 2228 (1995).
7. C. Benvenuti and P. Chiggiato, Vacuum **44**, 511 (1993).
8. C. Benvenuti, P. Chiggiato, F. Cicoira and Y. Aminot, J. Vac. Sci.Technol. A **16**, 148 (1998).
9. C. Benvenuti, J. M. Cazeneuve, P. Chiggiato, F. Ciciora, A. Escudeiro Santana, V. Johanek, V. Ruzinov, J. Fraxedas, Vacuum **53**, 219 (1999).
10. P. A. Anderson, Phys. Rev. **47**, 958 (1935).
11. R. Gomer, "Field Emission and Field Ionisation", American Institute of Physics (1993).
12. K. Tsukui, K. Endo, R. Hasunuma, T. Osaka, I. Ohdomari, N. Yagi, H. Aihara, J. Vac. Sci.Technol. A, **11** (5), 01 (1993) .
13. O. Gröbner, Vacuum **45**, 767 (1995).
14. J. M. Lafferty, J. Appl. Phys. **32**, 424 (1961).
15. J. Z. Chen, C. D. Suen and Y. H. Kuo, J. Vac. Sci. Technol. A **5**, 2373 (1987).
16. P. A. Redhead, J. Vac. Sci. Technol. **3**, 173 (1966).

17. J. C. Helmer and W. H. Hayward, Rev. Sci. Instrum. **37**, 1652 (1966).
18. F. Watanabe, J. Vac. Sci. Technol. A **11**, 1620 (1994).
19. C. Oshima and A. Otaku, J. Vac. Sci. Technol. A **12**, 3233 (1994).
20. K. Akimichi, T Tanaka, K Takeuchi, Y. Tuzi and I Arakawa, Vacuum **46**, Issues 8–10, 749 (1995).
21. J. P. Hobson, J. Vac. Sci. Technol. **1**, 1 (1964).
22. M. Bernardini, S. Braccini, R. De Salvo, A. Di Virgilio, A. Gaddi, A. Gennai, G. Genuini, A. Giazotto, G. Losurdo, H. B. Pan, A. Pasqualetti, D. Passuello, P. Popolizio, F. Raffaelli, G. Torelli, Z. Zhang, C. Bradaschia, R. Del Fabbro, I. Ferrante, F. Fidecaro, P. La Penna, S. Mancini, R. Poggiani, P. Narducci, A. Solina, and R. Valentini, J.Vac. Sci.Technol. A**16**, 188 (1998).
23. Y. Ishikawa and K. Odaka, Vacuum **41**, (7–9), 1995 (1990).
24. G. Grosse and G. Messer, Proc 8th Int.Vacuum Congr. 399, (Cannes) (1980).
25. P. Marin, M Dalinas, G Lissolour, A Marraud, A Reboux, Vacuum **49**, 309 (1998).
26. J. K. Fremery, Vacuum **53**, 197 (1999).
27. S. Ichimura, K. Kokobun, M. Hirata, S. Tsukahara, K. Saito and Y. Ikeda, Vacuum **53**, 291 (1999).
28. H. C. Hseuh, M. Mapes, R. Todd, D. Weiss, AIP Conf. Proc., 671, 292 (2003).
29. H. Ishimaru, J. Vac. Sci. Technol. **A 7**, 2439 (1989).
30. H. Kurisu, T. Muranaka, N. Wada, S. Yamamoto, M. Matsuura and M. Hesaka, J. Vac. Sci. Technol. A **21,** L10–L12 (2003).
31. F. Watanabe, J. Vac. Sci. Technol. A **22**, 181 (2004).
32. M. Stutzman, AVS 53rd International Symposium, Nov-2006.
33. R. Calder, A. Grillot, F. Le Normand and A. Mathewson, 7th International Vacuum Congress, Vienna, Austria, 1977, edited by R. Dobrozemsky, F. Ruedenauer, F. P. Viehboeck and A. Breth (Technische Universität Wien, Vienna), (1977).

12

Applications

12.1 General

The mechanical force of the atmospheric air that exerts on the surface of an object, pushes the object in the direction of vacuum. Commonly known as 'suction', this principle has been widely utilized for medical applications and in industry for vacuum transportation of materials. Vacuum wafer transport technology is employed in tool automation applications in semiconductor wafer processing and other complex manufacturing environments. Vacuum transport is also utilized for movement of abrasive bulk materials, dust, granulates, slurries and lubricants. Vacuum transport of garbage is executed using a stainless steel pipe system into collection tanks where the garbage is intermediately stored until it is disposed by the tanker. Vacuum collection, also known as pneumatic waste collection, provides centralized collection of different types of wastes in an area marked by a continuous waste generation at many adjacent points at the same time. Employment of such systems is prompted by the comfortable and hygienic way in which waste can be transported away from the place where it has been generated. Vacuum packing of waste in strong bags can considerably reduce volume of the waste. Vacuum packing involves removal of air from the package prior to sealing after the items are placed in a plastic film bag. Vacuum-sealed foods last much longer compared to conventional storage methods. As a result, foods maintain their texture and appearance much longer.

Thermal processes can be much faster, cleaner and energy efficient if operated in vacuum environment. Such processes include vacuum evaporation, distillation, filtration, freeze-drying, melting, deposition, welding, brazing, and sintering. Enhanced evaporation of fluids can occur at lower temperatures under vacuum. Absence of chemically active gases in vacuum facilitates contamination-free, clean processes.

Excellent thermal and electrical insulating properties of vacuum are utilized in laboratories and industry. Certain particle accelerators employ vacuum to isolate cold regions from heat as in the case of the Dewar flask. High electric currents of the order of several kilo-amperes can be satisfactorily interrupted in a vacuum interrupter.

The minimal presence of gas particles in vacuum facilitates collision-free paths for charged particles in electron tubes and particle accelerators. Lower gas impingement rate in UHV and XHV conditions causes much lower contamination of solid surfaces; thus the surface studies can be made in the cleaner condition. The characterization of solid surfaces is important to understand various phenomena that take place at the surfaces and in close proximity of the surfaces. Surface analytical instruments supported by high and ultrahigh vacuum systems are employed for studies involving interactions at solid surfaces.

Selected few applications of vacuum are discussed here.

12.2 Surface Analysis

The physical and chemical characteristics of solid surfaces can be investigated by using surface analytical techniques. Various applications of these vacuum based techniques include study of nanomaterials, solar cells, thin films, micro- and nano-electronics, corrosion/oxidation of metals, catalysis and surface treatments. Major surface analytical techniques are discussed here.

12.2.1 Ion Scattering Spectroscopy

The ion scattering that occurs in different ranges of incident ion energy has been discussed in Section 3.2.1.1.

Equation (3.33) shows that the surface atoms can be identified from knowledge of the mass of the primary ion and energies of the primary and the scattered ions. In Ion Scattering Spectroscopy (ISS), the energy spectra of the back-scattered particles at a given scattering angle give direct identification of surface atoms in the first atomic layer. Further, controlled etching of the target surface using ion beam of higher current density yields a depth profile analysis in the form of the elemental composition of successive atomic sub-layers. This method is also known as Low Energy Ion Scattering Spectroscopy (LEIS). Rutherford Backscattering Spectroscopy (RBS)

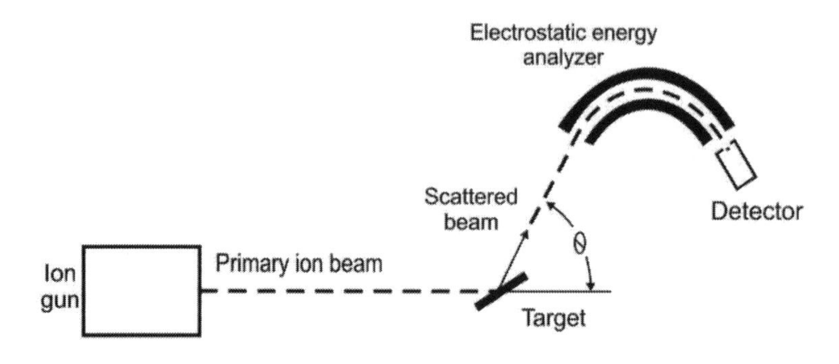

Fig. 12.1. Schematic diagram of ion scattering spectrometer.

generally corresponds to Medium Energy Ion Scattering (MEIS) with the incident ions in the range of 100 eV–200 keV. High Energy Ion Scattering corresponds to the incident ion energy in the range 1 MeV–several MeV.

From equation (3.33) one can see that E_1/E_0 is not a linear function of M^2. Thus separation of similar masses for heavier elements is difficult for masses above approximately 40 amu with He+, 60 amu with Ne+, and 80 amu with Ar+. The uncertainty of the inelastic losses and the neutralization rate depending on ion trajectories adversely affects quantification of surface analysis by using low energy primary ions.

ISS is extremely sensitive to the top surface layer or two monolayers (using grazing incidence). Composition of catalytic surfaces, thin film coatings, adhesion, arrangement of surface atoms, and localization of adsorbed atoms can be studied using ISS. This technique is capable of directly observing hydrogen atoms. ISS differs with the other ion scattering methods including MEIS, RBS and ERD, mainly in the energy of the primary ions. Figure 12.1 shows the schematic diagram of ISS.

Harrington et al[1] have examined the interface between Si and thermally grown SiO_2 by low energy He^+ ion scattering. Tongson et al[2] have used the ISS technique to optimize processing modes of precipitation of cobalt binder from solution onto WC grains for the development of WC–Co composite materials. Jacobs et al[3] have studied a series of Ni/Al_2O_3 catalysts prepared from the vapour phase by the atomic layer epitaxy. The activity measurements were combined with surface analysis by LEIS and X-ray Photoelectron Spectroscopy (XPS).

12.2.2 Secondary Ion Mass Spectrometry

The phenomenon of sputtering has been discussed in section 3.2.1.4. In Secondary Ion Mass Spectrometry (SIMS), a mono-energetic, singly charged, primary ion beam bombards the solid surface at incident energy between 1 and 30 keV. The intensity, energy, and orientation of the primary beam (relative to the sample) are controlled. The primary ions selected for this purpose can be argon or cesium or gallium. This results in release of monoatomic, polyatomic particles of the target material, re-sputtered primary ions, electrons and photons. The released particles have positive, negative, neutral charges with energies up to several hundred eV.

The energy distributions of the sputtered particles differ for atomic and molecular ions. Molecular ions have relatively narrow translational energy distributions while atomic ions have the entire kinetic energy in translational modes.

There is a wide variation in ion yields for various elements. The yields are sensitive to ionization potential for positive ions and to electron affinity for negative ions. Some of the emitted particles are charged positively or negatively and are known as secondary ions which are subsequently transmitted to a mass spectrometer.

Generally, the secondary ions are accelerated along a potential gradient before these are transferred into the mass spectrometer. The mass spectrometers utilize both magnetic and electrostatic analyzers, commonly referred to as sectors. The electrostatic filter reduces the energy range of the secondary ions so that they can then be separated into independent ion beams (based on the charge/mass ratio) by passing them through a magnetic field. Thus the multiple ion beams can be measured simultaneously. If the magnetic sector precedes the electrostatic sector, then mass resolution is improved at the cost of losing the ability to measure multiple ion beams simultaneously.

The secondary ions are analyzed by the mass analyzer to give the elemental, isotopic or molecular composition of the solid surface to a depth of 1 to 2 nm. With SIMS, the elemental detection limits range from parts per billion to parts per million. SIMS images can be obtained by scanning the primary ion beam across the sample surface and obtaining a spectrum. The depth profile can be obtained by sputter removal of sample layer-by-layer, which allows the species to be mapped in 3-dimensions. Other types of mass spectrometers that can be coupled to the SIMS source, include quadrupole and time-of-flight analyzers. Figure 12.2 shows the schematic of the secondary ion mass spectrometer.

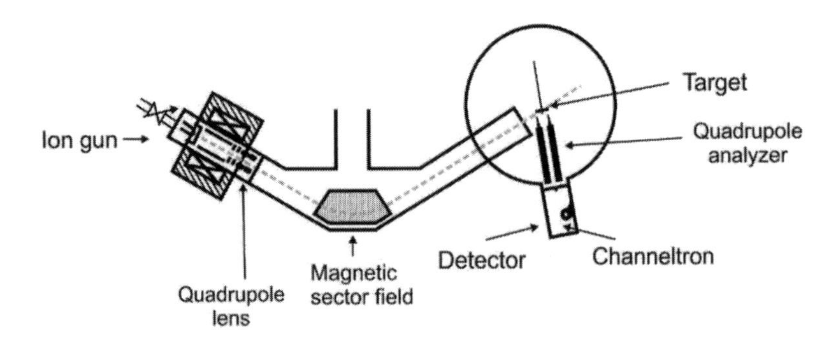

Fig. 12.2. Schematic diagram of secondary ion mass spectrometer.

Detection of elements from H to U is possible using SIMS. Concentration of elements down to 1 ppm can be detected. Isotopic ratios of accuracy 0.5% can be observed. The sputtered material can also include molecular species that makes the analysis difficult.

Benninghoven et al[4] have reviewed the theory, instrumental aspects and applications of SIMS. Francois-Saint-Cyr et al[5] implanted various elements such as Be, B, Na, Mg, Cl, K, Ca, Ti, V, Cr, Mn, Fe, Ni, Zn into silicon wafers as low dose impurities and annealed at different temperatures. Depth profiles were obtained by SIMS analysis to identify diffusion mechanism.

12.2.3 Auger Electron Spectroscopy

The Auger process has been discussed in Section 3.2.2.2. Auger Electron Spectroscopy (AES) uses a primary electron beam with energies in the range of several eV to 50 keV to excite the sample surface. The energy of the emitted Auger electron is characteristic of the element from which it is emitted. Detection and energy analysis of the emitted Auger electrons gives a spectrum of the Auger electron energy versus the relative abundance of electrons. Peaks in the spectrum correspond to the elemental composition of the sample surface. The chemical state of the surface atoms can also be determined from energy shifts and peak shapes. Figure 12.3 shows the schematic of the Auger electron spectrometer.

The Auger electrons have a relatively low kinetic energy. This limits their escape depth. The Auger electrons emitted from an interaction in the bulk below the surface lose energy by scattering reactions along its path to the surface. The Auger electrons emitted at a depth greater than about 2–3 nm do not possess sufficient energy to escape the surface and reach the detector. Thus, the analysis

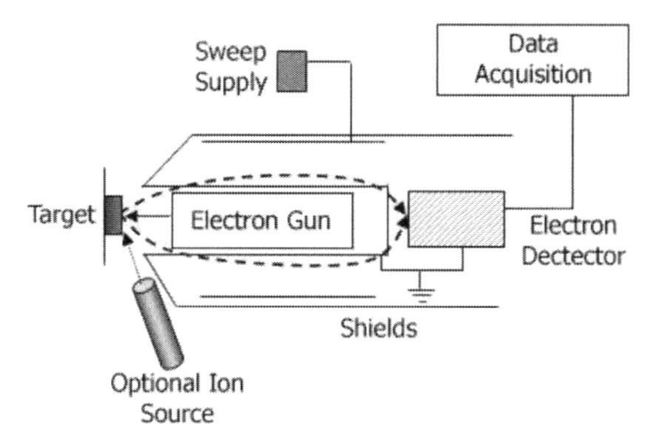

Fig. 12.3. Schematic diagram of Auger electron spectrometer (*Public Domain- Acarlso3 (Wikimedia Commons) https://commons.wikimedia.org/wiki/File:AES_Setup2.JPG*).

volume for AES extends only to a depth of about 2 nm. Analysis depth is not affected by the energy of the primary electron energy. The probability is greatest for the emission of an Auger electron for light elements. Analysis of the emitted secondary electrons is made and their kinetic energy is determined. The kinetic energy and intensity of the Auger peaks determine the identity and quantity of the elements. Auger electrons can be released only from the outer 5–50 Å of a solid surface at their characteristic energy thereby proving AES as an extremely surface sensitive method. A finely focused electron beam can be scanned to create secondary electron and Auger images. The electron beam can be positioned to perform microanalysis of specific sample features.

Applications of AES include materials characterization, failure analysis, thin film analysis, and particle identification for semiconductor and thin film head manufacturing. AES offers excellent spatial resolution (<1 μm) and surface sensitivity (~20 Å). Detection limits for most elements range from about 0.01% to 0.1%. Recent developments[6] in AES show that

- Spin polarization of Auger electrons can be used to study magnetized solid surfaces
- Results of resonant AES experiments provide information about femtosecond charge transfer dynamics
- Auger electron diffraction can be employed to determine surface structure
- Angle resolved AES can investigate excitation processes in solids

Rodekohr et al[7] used high resolution AES to determine both the surface and bulk composition of the high aspect ratio Sn whiskers grown from intrinsically stressed thin films (~6000 Å) of Sn on brass, deposited using cylindrical magnetron sputtering techniques. Results show that the whiskers are 100% Sn at the whisker base, shaft, tip, and up to a substantial depth into the whisker bulk. No evidence of pull-up from the brass substrate or surface contaminants is observed in the whiskers.

12.2.4 Electron Spectroscopy for Chemical Analysis

Electron Spectroscopy for Chemical Analysis (ESCA), also known as X-ray Photoelectron Spectroscopy (XPS), uses a focused monochromatic X-ray beam that irradiates the solid surface in UHV. Core electrons are emitted from the surface atoms by photoelectric effect. The emitted electrons have a kinetic energy E_k equal to the X-ray energy E_p less the binding energy E_b of the electron and the work function Φ.

$$E_k = E_p - \left(E_b + \Phi\right) \tag{12.1}$$

The detection of kinetic energies of the emitted electrons leads to identification of binding energies of the elements as the X-ray energy E_p is known. Different orbitals in the atom give rise to different peaks in the spectrum. The peak intensities are governed by the photoionization cross section. Extra peaks correspond to Auger emission. Energy spectra are observed as binding energy versus intensity. Peak intensities correspond to the quantitative elemental surface compositions. Figure 12.4 shows the schematic of the ESCA system. ESCA analysis covers the top 20–50 Å of the solid surface. This method can measure the surface elemental composition in parts per thousand range, chemical state and electronic state of the elements. The spatial resolution of 10 µm can be achieved with this method. XPS detects all elements. In practice, using typical laboratory-scale X-ray sources, ESCA can detect all elements with an atomic number (Z) of 3 (lithium) and above. It cannot easily detect hydrogen ($Z = 1$) or helium ($Z = 2$).

Using XPS, Chen and Burda[8] have investigated TiO_2 based nanometer sized photo-catalysts, including nitrogen doped TiO_2 nano-particles and have found evidence of O–Ti–N bond formation during the doping process. Moulder et al[9] have contributed on identification

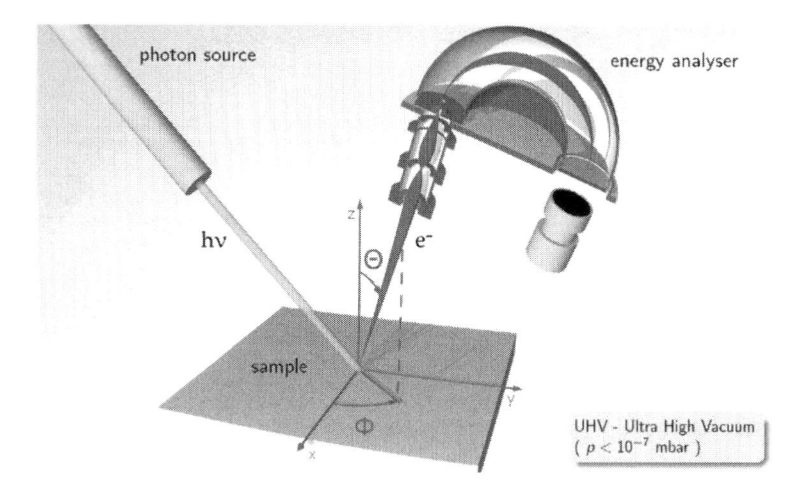

Fig. 12.4. Principle of angle-resolved photoelectron spectroscopy (*Public Domain- Saiht (Wikimedia Commons) https://commons.wikimedia.org/wiki/File:ARPESgeneral.png*).

and interpretation of the data. Synchrotron radiation can be used in place of X-rays as it is more intense and better collimated, offers tunability over a wide frequency range, and can produce extremely short photon pulses at a high frequency.

12.2.5 Low Energy Electron Diffraction and Reflected High Energy Electron Diffraction

In Low-Energy Electron Diffraction (LEED), the surface is bombarded with a collimated, 0.1 to 0.5 mm wide beam of low energy (approx. 10–200 eV) monochromatic electrons. Some of the electrons incident on the surface are backscattered elastically. A retarding field analyzer is used to detect these electrons as the inelastically scattered electrons are screened out. The relative position of the spots on the fluorescent screen can show the surface crystallographic structure if sufficient order exists on the surface. The diffracted spots will move as the energy of the incident electrons changes. LEED provides useful information about the crystalline solid surface. The analysis of the spot positions yields information on the size, symmetry and rotational alignment of the adsorbate unit cell with respect to the substrate unit cell. LEED can be used to study adsorption of gases on crystal surfaces. The schematic diagram of the LEED system is shown in Figure 12.5.

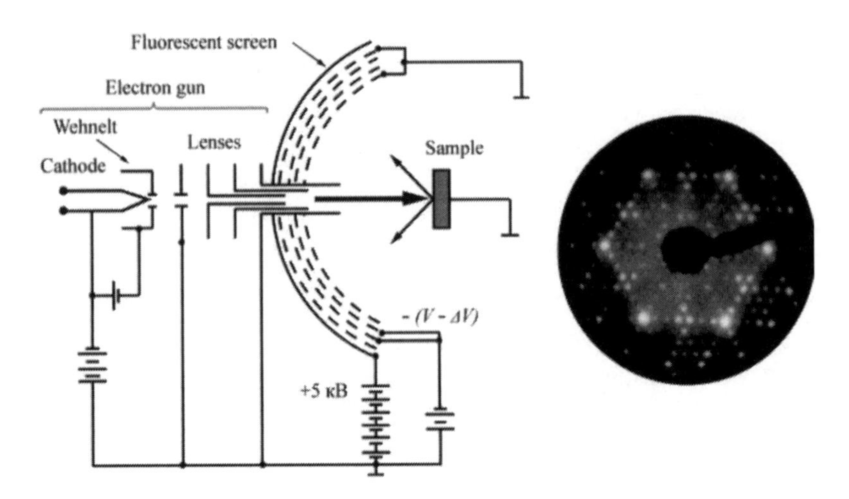

Fig. 12.5. Schematic diagram of LEED system and LEED image (*Reprinted from K. Oura: Surface Science, An Introduction, Springer eBook – Jan. 01 2003 with permission of Springer*).

The beam of electrons can be considered as a stream of electron waves incident normally on the surface. These waves are scattered by regions of high localized electron density, the surface atoms, which are considered to serve as scattering points. The range of the wavelengths of electrons employed in LEED experiments is comparable to the atomic spacing. Thus diffraction can be observed with atomic structure. Considering the backscattering of waves from two adjacent atoms at an angle θ to the surface normal, there is a 'path difference' (d) in the distance the waves have to travel from the scattering centers to a distant detector as shown in Fig. 12.6. The path difference ($a\sin\theta$) must be equal to an integral number of wavelengths for constructive interference to occur when the scattered beams eventually meet and interfere at the detector. For two isolated scattering centres the diffracted intensity varies between zero (complete destructive interference; $d = (n + ½)\lambda$) and the maximum value (complete constructive interference; $d = n\lambda$) With a large periodic array of scatterers, the diffracted intensity is only significant with the 'Bragg condition'.

$$a\sin\theta = n\lambda$$

$$d = a\sin\theta = n\lambda \tag{12.2}$$

where

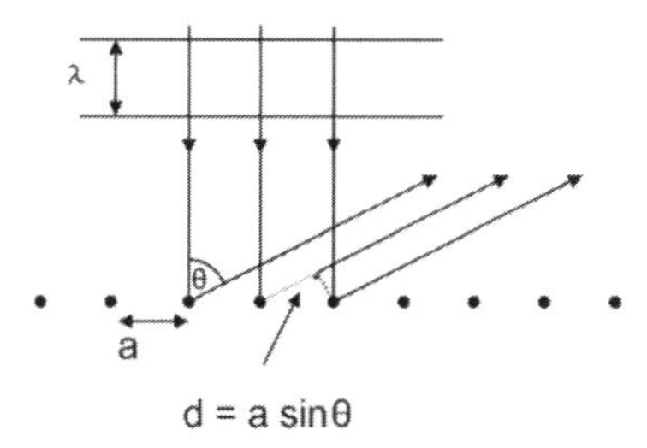

Fig. 12.6. Backscattering of waves from two adjacent atoms.

λ – electron wavelength; n – integer (..–1, 0, 1, 2,..); a – atomic separation.

The de Broglie relation is given by:

$$\lambda = \frac{h}{p} \tag{12.3}$$

where λ – electron wavelength; p – electron momentum; h – Planck's constant (6.62×10^{-34} s)

As

$$p = m \cdot v = (2mE_k)1/2 = (2m \cdot e \cdot V)^{1/2}$$

where m – the mass of electron (9.11×10^{-31} kg); v – velocity ($m \cdot s^{-1}$); E_k – kinetic energy; e – electronic charge (1.60×10^{-19} C); V – acceleration voltage

Thus, the electron wavelength,

$$\lambda = \frac{h}{(2meV)^{\frac{1}{2}}} \tag{12.4}$$

Oura et al[10] have discussed the modern surface analytical techniques including LEED.

In LEED, low energy electrons are used to provide large elastic scattering cross-section for back-scattered electrons and to keep the penetration depth of the electrons short.

Reflected High Energy Electron Diffraction (RHEED) employs large elastic scattering cross-sections of forward-scattered high energy electrons. The small penetration depth is achieved by using the grazing angle incidence of the incident electrons. In RHEED, a high-energy (5 to 100 keV) electron beam is incident on the surface at a very small angle relative to the sample surface. Incident electrons diffract from atoms at the surface of the sample, and a

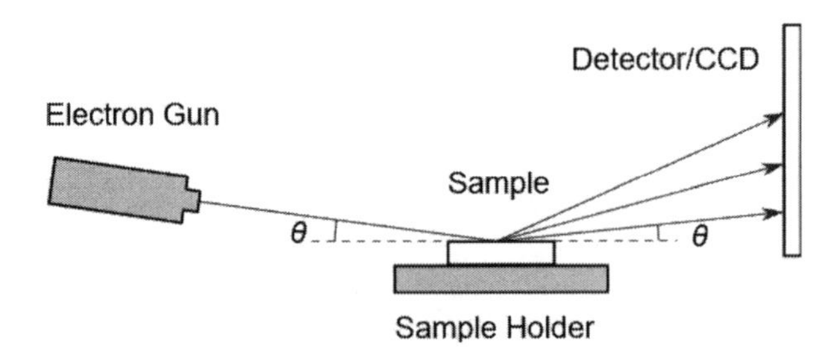

Fig. 12.7. Schematic diagram of the RHEED system (*CC BY-SA 3.0 -Atenrok (Wikimedia Commons) - https://commons.wikimedia.org/wiki/File:RHEED.svg*).

small fraction of the diffracted electrons interfere constructively at specific angles and form regular patterns on the detector. The electrons interfere according to the position of atoms on the sample surface, so the diffraction pattern at the detector is a characteristic of the sample surface. Figure 12.7 shows the schematic diagram of the RHEED system.

RHEED can be used to monitor the growth in MBE.

12.2.6 Electron Microscopy

In the electron microscope (EM) an electron beam illuminates the specimen and produces a magnified image. As electrons have wavelengths about 100 000 times shorter than visible light photons, EM can achieve resolution better than 50 pm and magnifications up to about 10^7X. The conventional visible light microscopes are limited by diffraction to a resolution of about 200 nm and useful magnifications below 2000X. EMs use the electron gun as the source of monochromatic electrons. The electrons released from the gun in the form of a beam are then focused by electrostatic and electromagnetic lenses, and are further transmitted through the specimen. The electrons emerging from the specimen provide information about the structure of the specimen that is magnified by the objective lens system of the microscope. The image can be viewed on a screen.

12.2.6.1 Scanning Electron Microscope

In the Scanning Electron Microscope (SEM), the electrons are focused into a fine probe. The schematic diagram of SEM is shown in Fig. 12.8. As the electrons penetrate the surface, some of the electrons that are emitted can be collected by suitable detectors. The image is

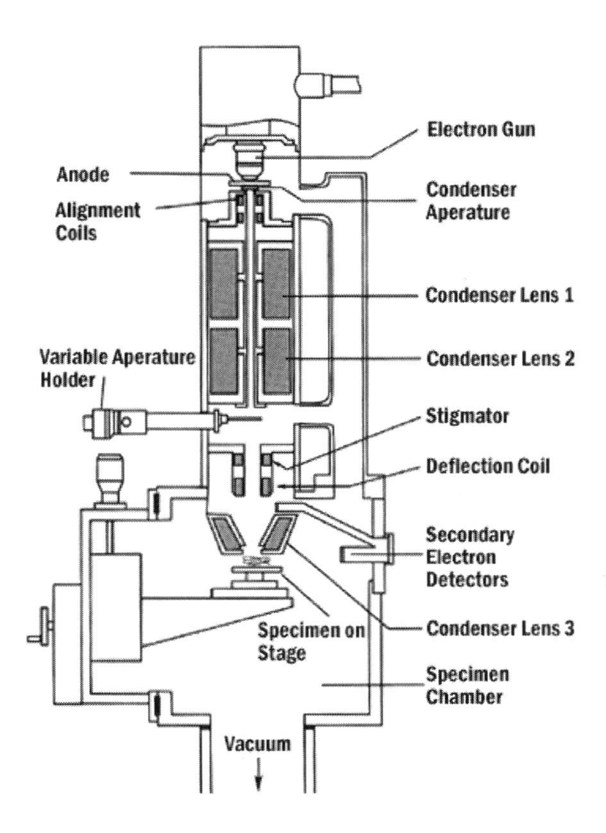

Fig. 12.8. Schematic diagram of the scanning electron microscope (*Central Microscopic Research Facility, Univ of Iowa. Reproduced with permission from University of Iowa*).

produced on the screen in such a way that each point on the screen corresponds to the point on the surface where the electron beam strikes.

The electron beam interacting with the specimen generates heat, emission of low-energy secondary electrons, high-energy backscattered electrons, light emission (cathodoluminescence) or X-ray emission. These provide signals with information about surface topography and composition of the sample.

The working voltage for SEM is between 2 to 50 kV and the diameter of the beam that scans the specimen is 5 nm–2 μm. The types of images generated by SEM include

- Secondary electron images
- Backscattered electron images
- Elemental X-ray images

The secondary and backscattered electrons are separated based on their energies. The secondary electrons have energies less than about 50 eV and the backscattered electrons possess energies greater than 50 eV. The back-scattered electrons that are reflected from the sample by elastic scattering and the characteristic X-rays released from the surface together reveal useful information including identification of the composition and measurement of the abundance of elements in the sample.

SEM is used only for surface images. The samples must be electrically conductive. Non-conductive materials are required to be carbon-coated. The materials with atomic number smaller than the carbon are not detected with SEM.

12.2.6.2 Transmission Electron Microscope

In the Transmission Electron Microscope (TEM) the electron beam

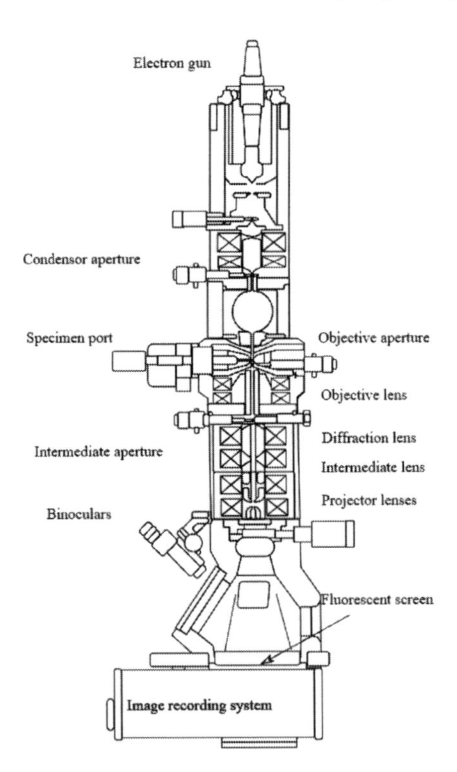

Fig. 12.9. Schematic diagram of transmission electron microscope (*CC BY-SA 3.0 -Gringer (Wikimedia Commons)- https://commons.wikimedia.org/wiki/File:Scheme_TEM_en.svg*).

interacts and passes through the specimen. Figure 12.9 shows schematic diagram of TEM. The electron beam, confined by the condenser lenses, further passes the condenser aperture and is incident on the sample surface. The electrons that are elastically scattered from the sample include the transmitted electrons, which pass through the objective lens. The objective lens forms the image display. Apertures are used to select the elastically scattered electrons that form the image. The beam later passes through the magnifying system which controls the magnification of the image and the projector lens. The formed image can be observed on a fluorescent screen and/or in monitor and can be printed on a photographic film.

12.2.6.3 Reflection Electron Microscope

In the Reflection Electron Microscope (REM), an electron beam is incident on a surface. The reflected beam of elastically scattered electrons is detected. This method is associated with reflection high energy electron diffraction (RHEED).

12.2.6.4 Scanning Transmission Electron Microscope

In the Scanning Transmission Electron Microscope (STEM) a thinned specimen is used to facilitate detection of electrons scattered through the specimen. The focused incident beam scans across the specimen. Thus, STEM provides high resolution of the TEM.

12.2.6.5 Field Emission Microscope

In the Field Emission Microscope (FEM) a metallic sample in the form of a sharp tip and a conducting fluorescent screen is mounted in ultrahigh vacuum. Figure 12.10 shows a schematic of FEM. The sample is held at a large negative potential (1–10 keV) relative to the fluorescent screen. The tip curve radius is ~100 nm. This results into an electric field near the tip to be of the order of 10^{10} V/m and causes field emission of electrons. Field emission of electrons is discussed in section 3.2.4.1. Electrons released from the tip travel toward the fluorescent screen where the image is formed. Bright and dark patches are formed on the fluorescent screen by the field electrons that travel along the field lines giving a one-to-one correspondence with the crystal planes of the hemispherical emitter. The emission current varies with the local work function based on the Fowler–Nordheim equation[11]. Thus, the projected work function map of the emitter surface is displayed in the FEM image.

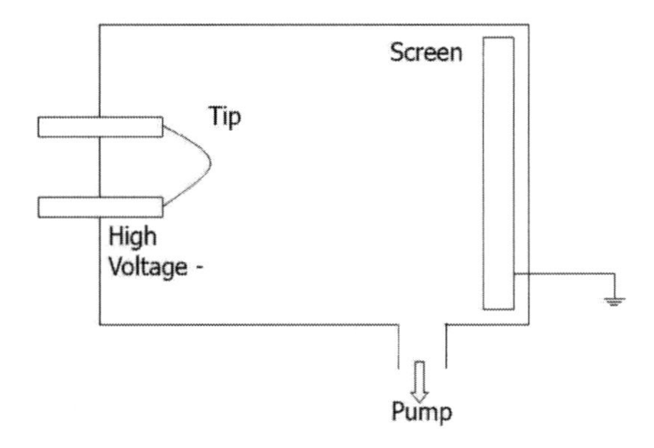

Fig. 12.10. Principle of field electron microscope *(CC BY-SA 3.0 - Creepin 475 (Wikimedia Commons)- https://commons.wikimedia.org/wiki/File:Field_emission_microscopy_(FEM),_experimental_set_up.jpg)*.

Linear magnifications of about 10^5 to 10^6 are attained. The spatial resolution of this technique is of the order of 2 nm and is limited by the momentum of the emitted electrons parallel to the tip surface, which is close to the maximum velocity (Fermi velocity) of the electron in metal.

Application of FEM is limited by the materials which can be fabricated in the shape of a sharp tip, and those can tolerate the high electrostatic fields. Thus, refractory metals with high melting temperature (e.g. W, Mo, Pt and Ir) are the appropriate objects for FEM. The magnification is proportional to the ratio of the radius of curvature of the screen on which the image forms to the radius of the metal tip. Magnification of the order of 10^6X can be achieved. The spatial resolution of FEM is of the order of 2 nm. The field emitter is required to be 'flashed' to clean by passing a current through a loop on which it is mounted. This allows the measurement of the variation of work function with orientation for different orientations on a single sample. The FEM has also been used to study adsorption and surface diffusion processes by making use of the work function change associated with the adsorption process.

Ehrlich and Hudda[12] have investigated the dependence of adsorption upon surface structure using FEM.

12.2.6.6 Field Ion Microscope

The Field Ion Microscope (FIM) was invented by Mueller[13]. FIM employs an object in the form of a sharp needle mounted on an electrically insulated stage, cooled to cryogenic temperatures (20

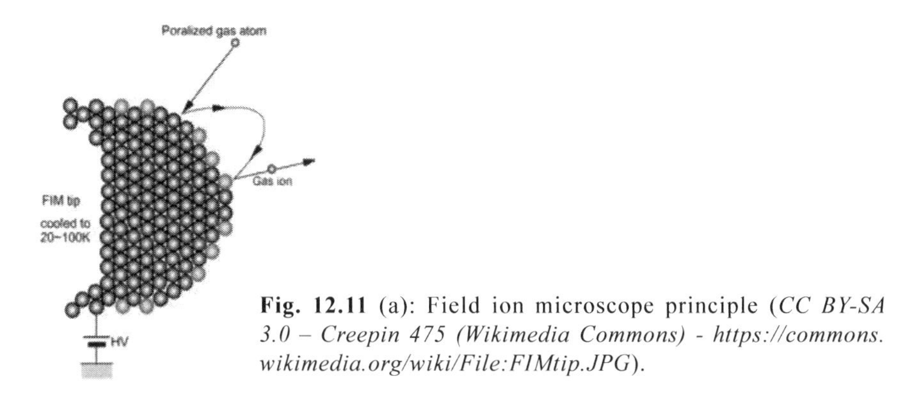

Fig. 12.11 (a): Field ion microscope principle (*CC BY-SA 3.0 – Creepin 475 (Wikimedia Commons) - https://commons.wikimedia.org/wiki/File:FIMtip.JPG*).

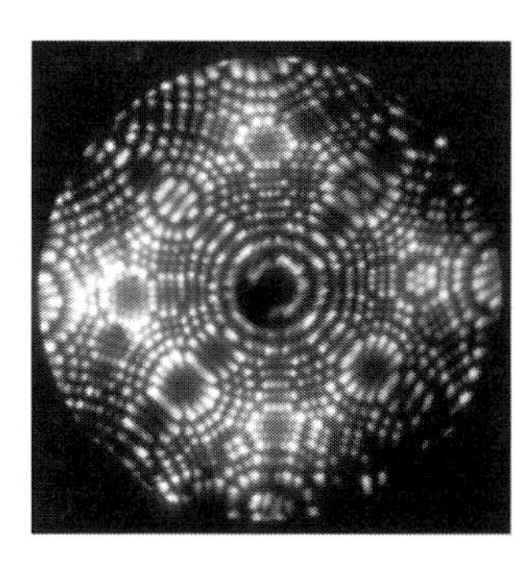

Fig. 12.11 (b): Field ion microscope image of the end of a sharp platinum needle. Each bright spot is a platinum atom. (*CC BY 2.0 – Tatsuo). Iwata (Wikimedia Commons) - https://commons.wikimedia.org/wiki/File:FIM-platinum.jpg*).

to 100 K) in the UHV chamber. The object is subjected to high electric field which causes polarization of the image gas atoms in the vicinity of the object. These gas atoms are attracted to the apex region of the object. The image gas atoms lose part of their kinetic energy resulting from collisions with the object atoms. During this process, the image gas atoms are thermally accommodated to the cryogenic temperature of the object. The image gas atoms are further field ionized by a quantum–mechanical tunnelling process due to the high electric field. The ions thus produced are repelled away from the surface in radial direction towards the micro-channel plate and the screen assembly positioned approximately 50 mm in front of the specimen. The micro-channel plate image intensifier positioned immediately in front of the phosphor screen produces about 100 electrons for each impinging ion. These electrons are accelerated towards the phosphor screen where they produce a visible image. The principle of FIM is illustrated in Fig. 12.11 *a*. A typical FIM image is shown in Fig. 12.11 *b*.

The controlled amount of the image gas is introduced into the vacuum system to produce a field ion image. The most common image gases are neon, helium, hydrogen and argon. The gas is ionized at the atom planes on the tip and produces an image that can have a magnification of up to 10^7 X. The FIM is used mainly to the study of metals and semiconductors. In the atom probe, individual atoms are released from the tip by application of pulsating electric field. The atoms travel through a time-of-flight type of mass spectrometer and their energy and charge-to-mass ratio are measured, thus facilitating determination of chemical nature of the atoms.

Ramanathan and Vijendran[14] studied the high field corrosion effect of nitrogen on tungsten, molybdenum and rhenium to evaluate its effect on the tip shape, size, and its crystallographic and chemical specificity.

12.3 Space Simulation

The major environmental characteristic of outer space is the vacuum. The gravitational attraction of planets and stars in space, brings gas molecules close to their surfaces thus leaving the space between almost empty. Few stray gas molecules between these bodies have their density extremely low. On the Earth, at sea level, the atmospheric pressure is 101 kiloPascals. The temperature at the sunlit side of objects in space at Earth's distance from the Sun can be over 120°C while the dark side can be at less than –100°C. The vacuum of intergalactic space contains a few hydrogen atoms/ions per cubic meter. At the surface of the Earth the air contains about 10^{25} molecules per cubic meter. Stars, planets and moons retain their atmospheres by gravitational attraction. Atmospheres do not have a clear boundary. The density of atmospheric gas gradually decreases with distance from the object until it becomes indistinguishable from that of the surrounding environment. The Earth's atmospheric pressure drops to about 3.2×10^{-2} Pa at 100 kilometers of altitude. Figure 12.12 shows variation of atmospheric pressure with altitude computed for 15°C at 0% relative humidity.

Vacuum is the most common condition encountered by spacecraft. Therefore, the study of the effects of vacuum on materials, components, or systems along with the other environments is important. Vacuum test chambers of many shapes, sizes, and capabilities have been used for the study. The research and development activities carried out in vacuum facilities has been crucial for progress in the space exploration program.

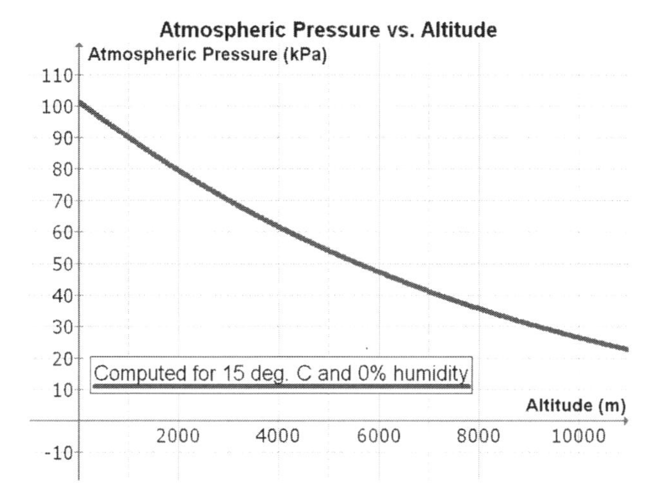

Fig. 12.12. Variation of atmospheric pressure with altitude (*CC0 1.0 CC0 1.0 CC0 1.0 CC0 1.0CC0 1.0CC0 1.0CC0 1.0 Public Domain - Geek.not.nerd (Wikimedia Commons): https://commons.wikimedia.org/wiki/File:Atmospheric_Pressure_vs._Altitude.png*).

Fig. 12.13. Simulation vacuum chamber of Space Power Facility of NASA (*Courtesy: NASA*).

The Space Power Facility[15] (SPF) of NASA houses the world's largest space environment simulation chamber, shown in Figure 12.13, measuring 100 ft. in diameter by 122 ft. high. The facility was designed and constructed to test both nuclear and non-nuclear space hardware in a simulated Low-Earth-Orbiting environment.

The facility can sustain a pressure of 10^{-4} Pa; simulate solar radiation via a 4-MW quartz heat lamp array, solar spectrum by a 400-kW arc lamp, and cold environments (–320°F) with a variable geometry cryogenic cold shroud. The concrete chamber enclosure that surrounds the vacuum chamber serves as a primary vacuum barrier from atmospheric pressure. The space between the concrete enclosure and the aluminum test chamber is pumped down to a pressure of 2.6×10^3 Pa during test.

The vacuum pumping system is equipped with five turbo-pumps of speed $2.2\ m^3 \cdot s^{-1}$, sixteen 48 inch diameter LN_2-baffled oil diffusion pumps of speed $700\ m^3 \cdot s^{-1}$, ten 48 inch diameter cryo-pumps of with vacuum isolation gate valves.

Pumpdown times are at the atmospheric pressure to 2.6×10^3 Pa: 2 hrs; 2.6×10^3 Pa to 10^{-1} Pa: 4 hrs; 10^{-1} Pa to 10^{-4} Pa: 2 – 6 hrs.

12.4 Gravitational Waves

The detection of gravitational waves[16] was made at the Laser Interferometer Gravitational Wave Observatory (LIGO) in September 2015. The two 4 km long tubes at 90° to one another and built on levelled ground serve as arms of a laser interferometer. An interference pattern formed at the detector reveals shifts in the period of the two light beams. The tubes are maintained in vacuum to avoid refraction of the laser beams and to eliminate sources of noise as the sound waves cannot propagate in vacuum, thus preventing the possible vibrations of the internal mirrors. 1100 hours of continuous pumping was necessary to achieve the required vacuum in the chambers. The vacuum was achieved by employing combination of turbo-pumps, SIPs, cryo-pumps and bake-out of the chambers

Data from the gravitational wave detector is expected to locate the sources of gravitational waves in the sky.

12.5 Vacuum Furnaces

Vacuum furnaces are employed for various applications such as brazing, sintering, heat treatment/annealing, degassing and carburizing. In a vacuum furnace the components assemblies to be thermally processed are surrounded by vacuum. The presence of vacuum minimizes heat transfer to the exterior. Other advantages include uniformity of the required high temperatures in the hot zone, capability of controlling temperature within a small area and absence of contamination of the items.

The vacuum furnace generally comprises

- Main chamber/tank
- Hot zone, which is the core of the furnace
- Vacuum pumping system
- Cooling system
- Power supply and controls

The main chamber/tank of the furnace is made of stainless steel and is generally cylindrical in shape. The loading port has a dished flange cover which can be opened for loading purpose. The other port of the tank is a closed flanged port and accommodates various electrical and water feedthroughs which are used for electrical connections to hot zone and for chilled water and argon gas circulation of the hot zone exterior. The parts to be thermally processed in vacuum are mounted in the hot zone located in the tank of the furnace. The shape and size of the hot zone is designed to accommodate the items. The heater elements are mounted on the internal walls of the hot zone. The design of the hot zone facilitates uniform heat influx to be incident on the items. The heater elements can be of different refractory materials such as molybdenum or graphite, depending upon the application and are suitably shaped and spaced in the hot zone. The hot zone is insulated thermally by the vacuum in the tank and by radiation shields of molybdenum and stainless steel to minimize the loss of heat from the hot zone to the exterior. In some furnace designs, the external walls of the hot zone assembly are continuously cooled during thermal processing, using chilled water circulating through copper tubes brazed on the walls. Thermocouples are mounted at various points in the hot zone to monitor the temperature.

Pumps used for the furnace mainly depend upon the volume of the furnace tank, the gassing loads and the working pressure. A combination of oil diffusion pump–rotary piston pump can be adequate for vacuum brazing furnace. Cryo-pumped and turbo-pumped vacuum furnaces can also be used.

Figure 12.14 shows a typical vacuum brazing furnace used for production application. The diffusion pumped vacuum system also includes a holding rotary mechanical pump, a refrigerated chevron baffle cum isolation valve and other valves for isolating different parts of the vacuum system. The valves are electro-pneumatically operated.

Fig. 12.14. Vacuum furnace (Courtesy: Cambridge Vacuum Engineering).

An elbow separates the vacuum system from the furnace tank and prevents pump oil vapors from directly entering the chamber. Vacuum furnaces that are required to handle large gas loads for processes such as vacuum sintering, utilize mechanical booster (Roots) pump, backed by an oil-sealed rotary pump, a combination which offers sufficient pumping speed in the pressure range where the diffusion pump and the oil-sealed rotary mechanical pump offer lower speeds.

A closed loop refrigeration system is used to circulate the LN_2 coolant around the exterior of the chevron baffle. In addition, chilled water is made to circulate around the diffusion pump, and the exterior of the hot zone. Heat exchanger system is employed to cool the furnace by circulating pressurized dry argon gas through the hot zone area. It serves a faster cooling of the furnace hot zone without causing any oxidation after the desired process in the furnace. A programmable system is employed to control the temperature, heat input, vacuum system operation, heat exchanger.

In a vacuum brazing furnace, processing of a number of components and assemblies involves initial pumpdown to a pressure of about 1 Pa, using the oil-sealed rotary mechanical pump while bypassing the diffusion pump and later using the diffusion pump with the backing pump for further evacuation to a pressure of about 1×10^{-3} Pa. If the furnace is not exposed directly to the atmospheric air and filled with dry argon or nitrogen gas during the previous cycle, the pumpdown is much faster and generally takes a few minutes time.

The next stage involves heating of the load in vacuum in a programmed manner. The temperature is made to rise in steps of 'ramps' and 'soaks' to ensure that the heat distribution is such that

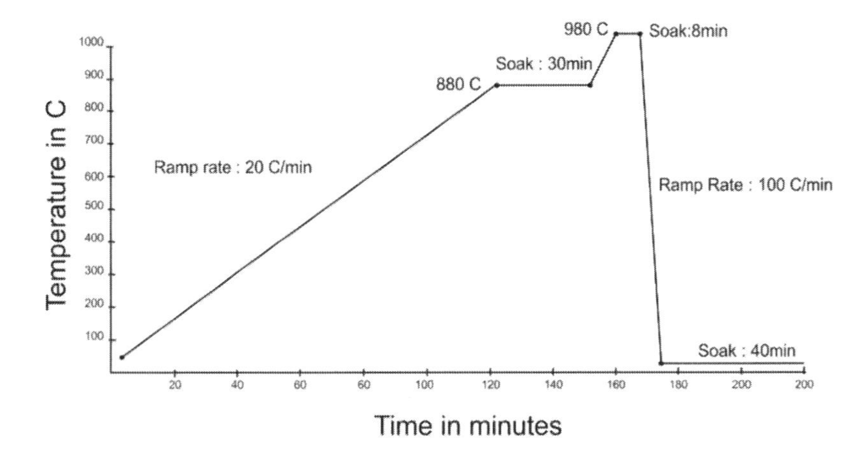

Fig. 12.15. A typical temperature-time thermal cycle of vacuum brazing furnace.

the load attains uniform temperatures. A typical temperature–time cycle of a vacuum brazing furnace is shown in Fig. 12.15.

The pressure in the furnace shows increase as the temperature is raised due to the outgassing of the load and the hot zone walls. Nevertheless, the pumping action of the furnace prevents the pressure to exceed beyond a safe limit. A sharp peak in the pressure level is observed when the temperature reaches the brazing temperature. This is due to the significant amount of gas released due to the melting action at the brazing temperature. Heat to the furnace is then cut-off and the furnace temperature starts falling. At this stage cooling by circulating gas takes over and the rate of fall in temperature is much faster. The forced cooling is terminated as the temperature nears the room temperature. The furnace door is then opened and the brazed components unloaded.

12.6 Particle Accelerators

The particle accelerators require vacuum in order to maintain the particles in the desired trajectories by minimizing their collisions with the residual gas. In the storage rings of the accelerators, UHV conditions are essential to achieve the beam life times in excess of 10 hours. Design of UHV systems for storage rings of particle accelerators is based on various parameters such as

- Lifetime of the beam for the full beam current
- Operating pressure and its uniformity across the storage ring
- Length, volume and surface area of the storage ring to be pumped
- Material of construction
- Surface processing and conditioning
- Leak detection
- Types and capacities of the pumps and their distribution across the storage ring
- Thermal outgassing from the walls
- Radiation and charge induced desorption loads from the walls
- Hardware such as valves, bellows, flanges, gaskets, mechanical manipulators, electrical feedthroughs
- Total and partial pressure measurement gauges

The lifetime is the time period to reduce the beam particles to $1/e$ of the initial value. Contribution to the lifetime can be from elastic and inelastic collisions of the charged particles with residual gas and mutual interactions of the charged particles.

The 26.7 km long Large Hadron Collider (LHC) at European Organization for Nuclear Research (CERN) is considered as the world's largest vacuum system[17]. It has separate vacuum systems for the beam pipes, for thermally insulating the cryogenically cooled magnets and for thermally insulating the helium distribution line. It operates at different levels of pressure. Figure 12.16 shows the LHC. It has 54 km long UHV system for the circulating beams in two superconducting storage rings with 50 km of insulation vacuum at a pressure of about 10^{-2} Pa. 48 km out of 54 km length is maintained at 1.9 K while the 6 km of straight sections, at room temperature employ NEG coatings. The UHV system utilizes 780 SIPs, 170 BA gauges, 1084 gauges including Penning and thermocouple gauges. The beam lifetime is 100 hours. The performance of the LHC vacuum system is governed by distributed pumping including NEG coatings in the ambient temperature sectors and cryo-pumping in the cold sectors.

The initial pumpdown time to attain a pressure of 10^{-4} Pa using TMP is about 2 to 3 weeks. Following this pumpdown, the components are baked at 230°C–320°C. The insulation vacuum is used in LHC to reduce the heat losses between the cold parts at liquid helium temperature and the external envelop at ambient temperature. The LHC is made up of eight 2.3 km-long arcs and eight 1.1 km-long straight sections. The arcs are maintained at 1.9 K using

Fig. 12.16. Large Hadron Collider (LHC) at CERN (Courtesy: CERN).

cryo-pumps to operate the LHC's superconducting bending magnets. The straight sections at room temperature house equipment such as beam injectors, extractors, collimators and particle detectors. The straight section vacuum monitored by 1500 pressure gauges, employs non-evaporable getter (NEG) coatings which serve as distributed pumping system, effective for removing all gases except methane and the noble gases. These residual gases are removed by 780 ion pumps. The room-temperature sections allow 'bakeout' of all components at 300°C.

As proton beams circulate in the LHC, the pressure increases by up to five orders of magnitude due to desorption caused by the electrons emitted from the inner surface of the beam pipe. These electrons, caused by stray protons, are accelerated towards the opposite wall of the beam pipe by the proton beam's positive charge, to release secondary electrons. The electrons generate negative space charge and perturb the circulation of the protons. The proton beam then diverges and protons are lost into the beam-pipe walls. The local increase in pressure is more likely to cause a high-energy proton to get lost to the superconducting magnets which can suddenly go from a superconducting to a normal conducting mode. This phenomenon is a potential limitation for the operation of the LHC at high energy and intensity. To minimize desorption, the beam pipes are coated with materials that offer much lower secondary electron yield. The situation in the straight sections is much less critical because of the application of NEG coatings, which have a low electron-multiplication effect causing suppression of the electron and the ion avalanches.

12.7 Plasma Applications

12.7.1 Plasma Fusion Machines

Nuclear fusion involving magnetic confinement of plasma is considered as the most promising process for generation of electric power. The deuterium–tritium fusion reaction that produces energy is described in the equation 4.13.

$$_1H^2 + {}_1H^3 \rightarrow {}_2He^4 + {}_0n^1 + 17.6 \text{ MeV}$$

Tokamak arrangement uses a magnetic field to confine plasma in the shape of a torus. The magnetic field lines move around the torus in a helical shape to achieve stable plasma equilibrium. A helical field can be generated by combining toroidal and poloidal fields. The toroidal field is produced by electromagnets surrounding the torus, and the poloidal field is produced by the toroidal electric current inside the plasma that is induced by electromagnets. *The International Thermonuclear Experimental Reactor* (ITER) is a thermonuclear fusion device of the Tokamak type, based on a deuterium–tritium plasma operating at temperature over 100 million K, to produce 500 MW of fusion power.

Day et al[18] have discussed the ITER vacuum systems. The main pumping systems of ITER include six cryo-pumps for the torus (1350 m^3), four cryo-pumps for the neutral beam injection systems (570 m^3) used for plasma heating and two cryo-pumps for the cryostat (8400 m^3) that contains the superconducting magnets. These will be cooled with supercritical helium. The vacuum vessel is a hermetically-sealed steel container inside the cryostat that houses the fusion reaction and serves as safety containment barrier. The plasma particles spiral around continuously without touching the walls in the doughnut-shaped chamber, or the torus.

The vacuum vessel will have double walled steel enclosure to facilitate circulation of cooling water. The inner surfaces of the vessel will be covered with blanket modules that will provide shielding from the high-energy neutrons produced by the fusion reactions. The ITER vacuum vessel will have an internal diameter of 6 meters, width of about 19 meters, height of 11 metres, and weight approximately 8,000 tons. Figure 12.17 illustrates a section of the ITER vacuum chamber.

For roughing pump systems, Roots pumps are used as the first two stages in all the pumps. One option uses a screw pump to back

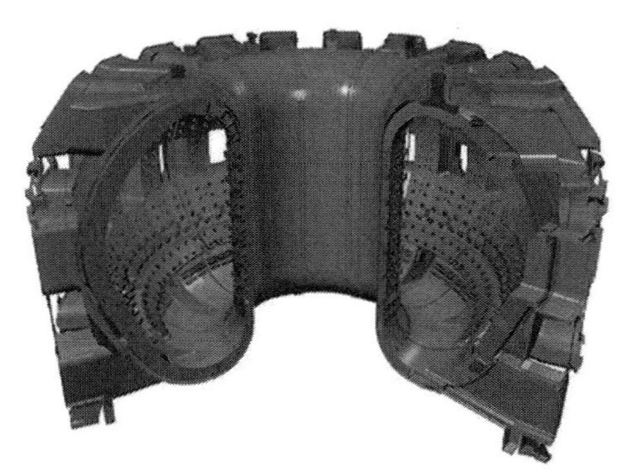

Fig. 12.17. Sectional view of vacuum chamber of the ITER Tokamak (*Courtesy: ITER*).

the Roots pumps. Pumping speed of 5000 m^3/h is possible with this configuration. Mechanical pumps and cryogenic pumps evacuate the vessel and the cryostat until the pressure inside has dropped to 1 to 10 Pa. Considering the volume of ITER, this operation is expected to take 24 to 48 hours.

Cryogenic pumping is employed primarily to handle high gas throughputs. The vacuum systems are designed also to withstand disruptions that can lead to high loads, radiation, and are required to be compatible with handling of tritium. Spray-coated charcoal cryo-panels, cooled by 4.5 K supercritical helium at 0.4 MPa are used in high-vacuum pumping systems of ITER. The charcoal used is microporous, granular, highly activated carbon, derived from coconut shells. The cryo-pumps have a 4.5 K cryo-sorption panel system and a 80 K thermal shielding and inlet baffle system to reduce thermal loads on the 4.5 K system. The torus cryo-pumps have 28 cryo-sorption panels. The required integral pumping speeds are 3800 $m^3 \cdot s^{-1}$ for H_2, and 2600 $m^3 \cdot s^{-1}$ for D_2.

Day[19] has described a three-stage cryo-pump developed for space simulation chambers requiring a pressure of about 10^{-4} Pa. The pump inlet baffle is cooled to 77 K using liquid nitrogen. The pump interior has a two-stage cooler with a 50 K condensation stage and a 20 K charcoal covered sorption stage. The pump has a nominal pumping speed of 50 $m^3 \cdot s^{-1}$ for nitrogen.

12.7.2 Vacuum Plasma Spraying

Vacuum plasma spraying (VPS) involves injection of metal or ceramic powder (10–50 μm) into a hot gas plasma (~10 000 K) that causes melting of the injected material. The molten droplets are projected at high velocity onto a substrate to form a coating. The VPS spray chamber is filled with inert gas and maintained at low pressure (~10^4 Pa) after evacuation during spraying to facilitate coatings of reactive materials to be produced without oxidation. The high droplet impact velocities and low oxidation associated with vacuum allow ~100% dense coatings to be formed.

Vacuum plasma spraying is preferred to atmospheric plasma spraying (APS) for oxidation sensitive materials such as Ti alloys, and/or where improved adhesion and density is required. Thomas et al[20] have investigated the effects of the substrate/coating interface 3D geometry on stress distributions using finite element analysis and VPS experiments to manufacture up to 2 mm thick W coatings.

12.8 Thin Films Applications

Thin films deposited in vacuum have several applications in the fields including optics, electronics, metallurgy, magnetics, mechanics and photovoltaics.

Chemical vapour deposition[21] and plasma chemical vapour deposition[22] (which also needs vacuum for producing plasma) are potentially promising techniques for thin film coatings. Excluding the coatings on architectural glass, most films are presently deposited in vacuum.

Antireflection coatings[23] on lenses are used in a number of optical objects such as eyeglasses, cameras and in optical equipment including telescopes, microscopes. Such coatings generally consist of layers of contrasting refractive indexes. The thicknesses of the alternate layers are selected to produce destructive interference in the reflected beams from the interfaces, and constructive interference in the transmitted beams.

High-reflection (HR) coatings are based on the periodic layer system with two materials, one with a high index, and the other with low index material. Such a system enhances the reflectivity of the surface in a certain wavelength range known as band-stop. The width of the band stop is decided by the ratio of the two indices for quarter-wave system. The maximum reflectivity increases up to about 100% with a number of layers stacked. The reflected beams interfere constructively with one another to maximize reflection and minimize transmission. Common HR coatings can achieve 99.9% reflectivity over a broad wavelength range.

Solar cells and solar modules directly convert the solar radiation into electricity. A photovoltaic absorber material is required for the solar cell. The material needs to absorb the incoming light efficiently and produce mobile charge carriers, electrons and holes that are separated at the terminals of the device without significant loss of energy.

A thin-film solar cell is made by depositing one or more thin layers of photovoltaic material on a substrate, such as glass, plastic or metal. Thin film cells are flexible and lightweight. Rigid thin film solar panels that are sandwiched between two panes of glass are employed in some photovoltaic power stations. Cadmium telluride (CdTe) is mainly used as the semiconducting material in thin film photovoltaics. It is sandwiched with cadmium sulfide to form a *p–n* junction solar photovoltaic cell. Some of the world's largest photovoltaic power stations employ CdTe photovoltaics. Other materials used for solar cells include copper indium gallium selenide and amorphous silicon. Sharma et al[24] have discussed the subject in their review article.

Thin-film interference is the optical phenomenon responsible for the coloured reflections that we observe. It occurs in structures composed of one or more transparent thin films, having typical thickness similar to the wavelength of light. Thin films are used for decorative applications. Reflection and refraction occurs at the top interface for the light incident on a single transparent thin film. Light transmitted into the film travels till it reaches the bottom interface, where a part is reflected and a part is transmitted. This process repeats as several times. Due to the wave nature of light, each portion of it can be imagined as a partial wave with its own wavelength, amplitude and phase. The manner in which the partial waves interfere determines the wavelengths that are transmitted and those reflected. The amount of the optical phase accumulated by a partial wave during the transit through the film depends on 'optical thickness' which is governed by the thickness of the layer, the incidence angle and the refractive index, thus giving rise to the destructive and constructive interference condition and to the emerging colours.

In the compact disc (CD) manufacturing process, the developed glass master is deposited with a thin film of nickel up to a typical thickness of around 400 nm. In the computer hard disc manufacturing process, the developed glass master is coated with a thin film of nickel of thickness about 400 nm in vacuum. During the replication process, the discs are passed through the metallizer where the discs are plasma sputter coated with aluminum alloy.

Thin film deposition is one of the methods used for surface hardening[25] of steel and cutting tools. Thin films and coatings find application in a broad range of industrial products. These include coatings that provide antistatic properties, corrosion resistance, reduce wear, and promote adhesion.

Thin films find immense applications in the field of microelectronics where conductive, resistive, and/or insulating films are deposited/sputtered on a ceramic or other insulating substrate. The films are deposited, photoprocessed and etched to form the required pattern. Thin film IC technology involves thin film deposition of components and interconnections on a glass or ceramic substrate. Thin film components such as resistors and capacitors produced by thin film techniques can also be used as discrete devices offering superior performance. Chopra[26] has discussed microelectronics applications of thin films.

12.9 Vacuum Interrupters

The vacuum interrupter (VI) is one of the most stringent applications of vacuum technology which involves high voltages up to 72 kV, high currents up to 63 kA, high arc temperatures and strong mechanical impacts. VI is a vacuum sealed-off switching device which forms an integral part of vacuum switchgear equipment such as vacuum circuit breaker and vacuum contactor. The major function of VIs is to make or break electric currents satisfactorily in vacuum environment. The excellent dielectric properties of vacuum are utilized in design and construction of VIs. A gap of 100 mm in vacuum breaks down at 800 kV. Figure 12.18 illustrates a typical VI.

A pair of electrical contacts are enclosed within high-alumina ceramic envelop and are linked with OFHC copper electrodes projecting out of the envelop. One contact is fixed and the other contact is movable. The contact's axial movement is facilitated by stainless steel bellows linked to the end plate. A coaxial metal shield surrounding the contacts is located closer to the internal surface of the envelop to prevent metal vapors from condensing on the internal surface. Electrostatic shields are placed at both the ends. A shield surrounds the bellows for its protection. The entire assembly is sealed off in high vacuum at about 10^{-4} Pa in vacuum brazing furnace.

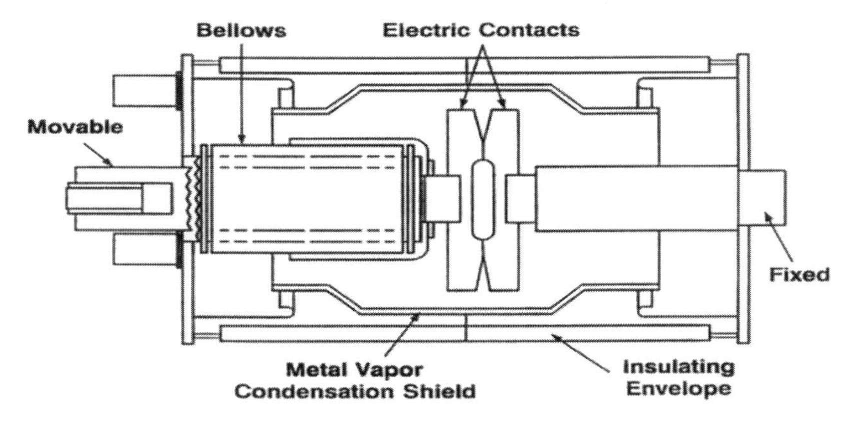

Fig. 12.18. Vacuum interrupter (*Courtesy: Eaton Corporation*).

The assembly of the components of VI is made in a dust-free and controlled environment clean room to avoid possible contamination.

The contacts are held tightly closed together in normal condition as the normal circuit current passes through them. When undesirable excessive or short-circuit current passes, the external mechanism that senses the increased current, forces open the contacts to the extent of a gap of a few mm to about 10 mm. As the contact parts, an arc comprising metal vapor plasma is formed between the contacts. The arc gets extinguished at the current zero due to absence of plasma formation. The VI thus transforms from an electrical conductor to an insulator and protection to the electric circuit from causing damage is achieved.

Major technologies that are involved in VI manufacturing include, vacuum metallurgy, vacuum brazing and ceramic metallizing. Special materials and components that are used include, high-alumina ceramic, brazing alloys, OFHC copper, stainless steel bellows, NEGs. All components of the VI are thoroughly cleaned prior to their assembly.

The electrical contacts of VIs are the key components for reliable and satisfactory performance of VIs. Selection of materials for the contacts and the configuration of the contacts is mainly governed by the application. Slade[27] has assessed important material properties that are required to be considered for the contacts. These include the gas content, melting point, vapor pressure, electron emission, electrical and thermal conductivities, gettering action of the residual gases and structural quality. He has further discussed performance and application of the VI contact materials. Various materials for the VI contacts assessed by him include Cu–Cr with and without additives, Ag–WC, Cu–W with and without additives and Cu alloys. The shape of the contacts influences the motion of the vacuum arc across the contacts and within the gap between the open contacts.

Mass spectrometer leak detector is utilized to detect and locate leaks, if any. One of the methods that is used for testing the vacuum integrity of the sealed-off VIs is magnetron and inverted magnetron tests[28,29] in which cross electric and magnetic fields are applied to the VI and the current is measured as a function of pressure inside the VI. The current in the VI is identified as a series current of an inverted magnetron and of a magnetron through the central shield as a common electrode. In another test, the sealed-off VIs are stored for a few days in pressurized vessels to check the rise in pressure, if any, arising from the leaks. The sealed-off VI has a shelf life of over 30 years. Slade[27] has calculated the maximum leak rates for VIs to achieve a given life-time for different internal volumes of VIs. He has shown that with a maximum leak rate of 4.2×10^{-10} Pa · liters · s^{-1} a VI of internal volume of 4 liters will take 30 years to reach the internal pressure of 10^{-1} Pa which is considered as the maximum internal pressure for VI to operate satisfactorily.

12.10 Recent Advances

M. Mozetič et al[30] have discussed recent advances in vacuum science and applications. These include review of new developments in electron transport near solid surfaces, progress in analysis of carbon-based nanostructures and applications of surface analytical methods in conservation technology of cultural heritage objects. They have further reviewed use of UHV based analytical instruments for research, development and quality control of bio-interfaces. They have also discussed surface engineering that deals with the materials science and technology of modifying and improving the surface properties of materials for protection. Also reviewed by them is the synthesis of multifunctional surfaces, including active and adaptive control, design of interfaces, and implementation of modelling approaches based on density functional theory, molecular dynamics simulations, and finite element methods for materials development, engineered multilayers and nanostructured coatings. They have discussed recent progress in thin films including the subjects of growth of thin film on granulates and hollow microspheres, MAX phases, magnetic and oxide thin films, and nanostructured thin films, carrier mobility in organic thin films, quantum dot (QD) self-assembly, and super-lattices investigation using grazing incidence small angle x-ray scattering (GISAXS). They have further discussed topics in plasma science that include plasma nanoscience and plasma biomedicine.

References

1. W. L. Harrington, R. E. Honig, A. V. Goodman and R. E. Williams, Appl. Phys. Lett. **27**, (12) 165 (1975).
2. L. L. Tongson, J.V. Biggers, J. M. Bind, G. O. Dayton and B. E. Knox, J. Vac. Sci. Technol. **15** (3), 1129 (1978).
3. J. P. Jacobs, L. P. Lindfors, J. G. H. Reintjes, O. Jylha and H. H. Brongersma, Catal. Lett. **25**, 315 (1994).
4. A. Benninghoven, F. G. Rüdenauer, and H. W. Werner, "Secondary Ion Mass Spectrometry: Basic Concepts, Instrumental Aspects, Applications and Trends", Wiley, New York, (1987).
5. H. Francois-Saint-Cyr, E. Anoshkina, F. Steviea, L. Chow, K. Richardson, D. Zhou, J. Vac. Sci. Technol. B **19** (5), 1769 (2001).
6. R. P. Gunawardane, C. R. Arumainayagam, "Handbook of Applied Solid State Spectroscopy", 451 Edit. D. D. Vij, Springer (2006).
7. C. L. Rodekohr, G. T. Flowers, J. C. Suhling, M. J. Bozack, IEEE Holm Conf. on Electrical Contacts, 27–29 Oct. (2008).
8. X. Chen and C. Burda, J. Phys. Chem. B **108** (40), 15446 (2004).
9. J. F. Moulder, J. Chastain and R. C. King, "Handbook of x-ray photoelectron spectroscopy : A reference book of standard spectra for identification and interpretation of XPS data", Pub:Physical Electronics (1995).
10. K. Oura, V. G. Lifshits, A. Saranin, A. V. Zotov and M. Katayama, "Surface Science-An Introduction", Springer, Berlin, Heidelberg, New York (2003).
11. R. H. Fowler, L. Nordheim Proc. Royal Soc. A **119** (781), 173 (1928).
12. G. Ehrlich and F. G. Hudda, J. Chem. Phys. **35**, 1421 (1961).
13. E. W. Müller, Zeitschrift für Physik **131**, 136 (1951).
14. D. Ramanathan and P. Vijendran, J. Phys. Colloques **47** (1986).
15. R. N. Sorge, NASA/TM—2013–217816.
16. B. P. Abbott et al, Phys. Rev. Lett. **116**, 061102 (2016)
17. J. M. Jimenez, Vacuum **84**, 2 (1) (2009).
18. C. Day, D. Murdoch and R. Pearce, Vacuum **83**, Issue 4, 773 (2008).
19. C. Day, "Basics and Applications of Cryopumps", Proc. CERN Accelerator School on Vacuum, Platja d'Aro, Spain, 241. May (2006).
20. G. Thomas, R. Vincent, G. Matthews, B. Dance and P. S. Grant, Materials Science and Engineering: A**477**, Issues 1–2, 35 (2008).
21. J. H. Park and T. S. Sudarshan-Ed. "Chemical Vapor Deposition", ASM International (1951).
22. S. Yugo, T. Kanai, T. Kimura and T. Muto, Appl. Phys. Lett. **58**, 1036 (1991).
23. H. K. Raut, V. A. Ganesh, A. S. Nair and S. Ramakrishna, Energy Environ. Sci. **4**, 3779 (2011).
24. S. Sharma, K. K. Jain, A. Sharma, Materials Sciences and Applications **6**, 1145 (2015).
25. M. J. Schneider and M. S. Chatterjee, Revised from "Introduction to Surface Hard-

ening of Steels, Heat Treating", Vol. **4**, ASM Handbook, ASM International, 259 (1991).
26. K. L. Chopra and I. Kaur, "Thin Films Device Applications", Plenum Press, New York, (1983).
27. P. G. Slade, "The Vacuum Interrupter, Theory, Design and Application", CRC Press, Taylor & Francis Group, Boca Raton–London–New York (2007).
28. K. Kageyama, J. Vac. Sci. Technol. A **1**, 1522 (1983).
29. K. Kageyama, J. Vac. Sci. Technol. A **1**, 1529 (1983).
30. M. Mozetič, K. Ostrikov, D. N. Ruzic, D. Curreli, U. Cvelbar, A. Vesel, G. Primc, M. Leisch, K. Jousten, O. B. Malyshev, J. H. Hendricks, L. Kövér, A. Tagliaferro, O. Conde, A. J. Silvestre, J. Giapintzakis, M. Buljan, N. Radić, G. Dražić, S. Bernstorff, H. Biederman, O. Kylián, J. Hanuš, S. Miloševič, A. Galtayries, P. Dietrich, W. Unger, M. Lehocky, V. Sedlarik, K. Stana-Kleinschek, A. Drmota-Petrič, J. J. Pireaux, J. W. Rogers and M. Anderle , Journal of Physics D: Appl. Physics **47**, Number 15 (2014).

Index

G

H

I

R

S